Bonfire Lessons

How to leverage the magic in the universe
to reach your goals and dreams

David Ruuskanen – 2004

Note for Librarians: a cataloguing record for this book that includes
Dewey Decimal Classification and US Library of Congress numbers is
available from the National Library of Canada. The complete cataloguing
record can be obtained from the National Library's online database at:
www.nlc-bnc.ca/amicus/index-e.html
ISBN 1-4120-4104-X
Printed in Victoria, BC, Canada

TRAFFORD

Offices in Canada, USA, Ireland, UK and Spain
This book was published *on-demand* in cooperation with Trafford
Publishing. On-demand publishing is a unique process and service of
making a book available for retail sale to the public taking advantage
of on-demand manufacturing and Internet marketing. On-demand
publishing includes promotions, retail sales, manufacturing, order
fulfilment, accounting and collecting
royalties on behalf of the author.
Book sales in Europe:
Trafford Publishing (UK) Ltd., Enterprise House, Wistaston Road
Business Centre, Wistaston Road, Crewe CW2 7RP UNITED KINGDOM
phone 01270 251 396 (local rate 0845 230 9601)
facsimile 01270 254 983; info.uk@trafford.com
Book sales for North America and international:
Trafford Publishing, 6E–2333 Government St.,
Victoria, BC V8T 4P4 CANADA
phone 250 383 6864 (toll-free 1 888 232 4444)
fax 250 383 6804; email to bookstore@trafford.com

www.trafford.com/robots/04-1911.html

10 9 8 7 6 5 4 3

This book is dedicated to my children as they were my inspiration. What started out a few years ago as a collection of ideas that I wanted share with them grew into a much bigger project.

Robbie and Joanna...I love you.

Acknowledgments

I wish to acknowledge the many authors, scientists and philosophers who have opened my mind to the mysteries of life. A special note of gratitude to Susan Yates who provided guidance and encouragement as the manuscript took shape. A huge thank you to: John Saville, Nora Lynch, Karen Hanna, John Gunning and Otto Schmidt, who gave me feedback on early versions. And a very special thank you to my wife Barbara Stroud Ruuskanen who constantly supported me as this project came to life. I love you.

Contents

Preface

Imagine living the life of your dreams. Think about how empowered you would feel as you set goals and achieved them. Bonfire Lessons describes our magical universe and enables the reader to harness this magic to reach for the possibilities. This book is targeted to those that have the courage to unlock their potential and dare to overcome the challenges and obstacles that lie before them.

As we go about our daily lives we are unaware of the dynamic and magical environment we live in. Occasionally we think beyond our current activities to visions of our goals and dreams, which often seem very distant. When we consider the challenge to achieve these goals our mind produces many obstacles, making it difficult for us to conceive of the possibility. We create a paradox. On the one hand we consider the *possibilities*, on the other hand, we actively participate in negative self-talk, which only serves to convince us that we don't have enough time, money, or talent to achieve the desired outcome.

What we don't fully realize is that our goals are achievable. We can prove this to ourselves through "hindsight." Most of us can look back on a moment in our lives where we achieved an intended goal. Perhaps it was a job promotion, graduation from school, a business deal, a specific athletic award or something more uniquely personal.

Many of these goals were achieved with minimal conscious effort and in other instances a tremendous amount of focus was applied. Furthermore, we have all experienced unrealized goals because our life interests changed or our focus wasn't keen enough to manifest the dream. Sometimes we lose sight of our accomplishments because our focus is constantly shifted forward. We always seem to crave for more. Take a moment to think about your accomplishments. You will realize that you have already reached many goals in your lifetime.

In this fictional story which is based on fact, a 79-year-old teacher named Uncle Jack opens our eyes to the magic that is available to all. The focus of this story is to examine the truly magical environment in which we live, and by this under-standing it gives us the knowledge and confidence to pursue and manifest our goals and dreams. In fact, Uncle Jack's goal is to share the acquired knowledge of his life to enable oth-ers to reach their potential. Uncle Jack's lifetime pursuit is to help us break down the complex into a simplified form, allowing his students to eliminate their blocks, which create obstacles between themselves and their goals and dreams.

To first understand how to tap into this magic, we need to understand a little about the universe we live in, its origin, and its make-up. Each successive chapter introduces a new thought that illustrates that we have the capability to shape our own destiny and to challenge our current thinking. These chapters explore such subjects as:

1) $E=mc^2$ and relativity: mathematical equations and concepts conceived by Albert Einstein in the early 1900's that have revolutionized current thinking of the universe we live in.

2) Quantum mechanics: which operates at the op-posite end of the size spectrum when compared to Einstein's relativity and $E=mc^2$.

3) Manifestation: it is the spark to create anything and everything. How does a thought turn into a material object or an action?

What we will find out is that the world we live in is much more than the three dimensions that we consciously experience on a daily basis. Magic exists in the universe and we need to be consciously aware of it.

Okay, perhaps you are thinking that this is interesting but so what? What does our understanding of the universe, $E=mc^2$, quantum mechanics, manifestation and our basic thinking mean to us on a daily basis? How can understanding these concepts affect our lives and our future?

This is the essence of this story. By understanding some of the basic concepts of our universe, quantum mechanics and $E=mc^2$ we can effect our future. For some people this will reinforce their current thinking. For others, this is a chance to pick up concepts that have been locked in the minds of brilliant scientists, physicists, astronomers and philosophers who have not been able to express their ideas in a manner that those of us in the general population can relate to. Do you even know what $E=mc^2$ means, let alone how understanding it can change the way you think? It is a tragedy that as humans we have evolved our thinking and our knowledge through the minds of brilliant people but this information hasn't been shared with the masses because of its supposed complexity.

This is Uncle Jack's mandate. This is what he wants to change. Uncle Jack is a mythical character who is a master of many esoteric subjects. The setting is summer, alongside the mighty Ottawa River, which provides an opportunity for the characters to be away from their day-to-day lives. Uncle Jack is a permanent resident along the river, while the other characters are people who enjoy their cottages seasonally. Uncle Jack illuminates the mysteries and power that are available to all. He does this each night in his bonfire lessons beside the water. He creates a safe environment for his audience to seek answers to their questions about life and the pursuit of goals and dreams. Welcome to summer alongside the Ottawa River.

Part 1

The Universe We Live In

1 *The Invitation*

To understand the heart and mind of a person, look not at what he has already achieved, but what he aspires to do.

— Kahlil Gibran

He sat beside the fire and gazed across the river watching intently as a heron took flight, its silhouette clearly outlined against the bright orange of the setting sun. The air was still, occasionally punctuated by the crack and hiss of the fire. Returning his attention to the task at hand he carefully addressed the last letter to Tom Armstrong and then placed it down beside the others, one each for siblings Brian and Carolyn Smith. He smiled and his inner wisdom told him that his invitations were already accepted.

Uncle Jack is an energetic, kind and engaging fellow who has entertained everyone in Pine Ridge for as long as they can remember. He is a true uncle to no one, however, he is a loved person who is called uncle by everyone. He is a mountain of a man who is 79 years young. Uncle Jack is a master of life whose enlightenment radiates to all that are around him.

He has spent most of his life doing the things that he wanted to do. To those that know him, he never seems like he's working, had a dull moment or fallen to the depths that people sometimes do. Uncle Jack has a mission and that is to spend time enjoying life and ensuring that he is making it enjoyable for others, giving them the tools they need to pursue the joys of life on their own.

By the time the summer residents had all arrived at their cottages in Pine Ridge, Uncle Jack had set up much of the communal activity areas: the rafts were floating invitingly in the river awaiting the water games to be played, the badminton/volleyball court was awaiting its teams and players, the firewood was collected and piled awaiting the campfire stories, songs and long discussions. Pine Ridge is a collection of cottages and homes beside the ever flowing Ottawa River where the river is a half a mile wide and the banks are covered in majestic pines. Uncle Jack is one of the few permanent residents of Pine Ridge, enjoying the splendors that nature provides year round.

The children who spent their summers at Pine Ridge never knew that they were learning in the school sense through their experiences with Uncle Jack. They thought they were having a great time doing what kids do...having fun playing games and asking questions and seeking guidance from the person they all respected. The adults cherished their evening bonfire chats with Uncle Jack because it provided a time to reflect and to think about the possibilities.

The evening bonfire discussions occurred when everyone was away from the typical daily life that ran from September to June in the cities and towns from which they came. They were on hot summer nights lying on the beach gazing up at the magical universe as they thought about future goals and dreams.

The pace of life at Pine Ridge was very different from the typical daily life. This afforded time to learn from Uncle Jack and to connect with nature. That connectedness took many forms; exploring, refreshing swims in the river, peacefully canoeing, quietly fishing, skipping stones, listening to the rain drops as they rhythmically hit the roof of the cottage while drifting off to sleep, absorbing the beautiful rays of sunshine, smelling the refreshing breezes, walking along the forest paths, lying in the hammock allowing the gentle wind to rock them to sleep.

No matter what they were doing, there was one thing that was evident to all; everyone's outlook on life was different in Pine Ridge. One thought differently, time passed by differently, and one had a completely different perspective on the events in life. The changed perspective was perhaps synonymous with what the astronauts experience when they go off into earth's orbit in the space shuttle. They have a very unique opportunity to have a different perspective on life. They have a chance to look back at this small planet, which we call *earth* and realize that *earth* is only one very small beautiful microcosm of the universe. And they have this broad view that despite the challenges that we face on earth, we collectively bring love and hope to the billions of people with whom we live. While in space, these same astronauts realize that their own difficulties or challenges in life are not nearly as significant as they once thought. Perhaps it is only when you have a different perspective on life that the problems or challenges disappear and the realm of possibility appears.

The BMW convertible was speeding down the highway, the driver's blonde hair streaming behind him. The mid-day sun beat down on the black leather interior as the rock music blared from the speakers that surrounded him. Tom's casual driving style was evident as his elbow rested on the window ledge as he directed the car with his left hand. All the while the fingers on his right hand snapped to the music and the words to the song on the stereo were sung at the top of his lungs. It was a picture of happiness and freedom.

Tom was taking an extended vacation from the drudgery of work. Since graduating from university nine years ago, he had been working 50-60 hour workweeks with only a couple of weeks off a year. Yesterday had been his last day of work for the next two months. As he cruised down the highway on the beautiful summer afternoon, thoughts of work seemed far away.

In his youth Tom spent every summer at Pine Ridge. His summers were filled with great adventure and fun. After graduating from university at the age of 22 he hadn't been to Pine Ridge much more than a few weekends.

Tom is a tall wiry fellow with curly blonde hair and an engaging personality. He is typical of sorts where people may be gone for a decade or more from Pine Ridge but happily return at some point later in life. The great thing about Pine Ridge is that you don't ever really leave it completely; its energy is always with you beckoning your return.

Tom's career had taken a turn for the worse. It wasn't that he wasn't performing it's just that his success was defined by the expectations of his superiors. Over the years in his professional life Tom demonstrated his capabilities time and time again. He worked hard and demanded a great deal from himself. Perhaps he demanded too much. Now at the age of 31, despite his successful career, he is struggling to meet the expectations he set for himself in life. While the money flowed in, it flowed out just as fast while he lived the high life: sports cars, downtown condo, dinners, skiing, golf, the finest designer suits etc. Life had been great. He had friends, girl-friends, and teammates with whom he played basketball in the winter and baseball in the summer. But what Tom craved was something more than money. What he was looking for was how to have it *all*. At some point he wished to be married and to start a family and his plans included owning a home in the country and perhaps owning a cottage in Pine Ridge. He wished to have a career that excited him, not to have a job that paid him a great deal but kept him in the office 50 to 60 hours a week. He did not want to be a slave, which is how his job was beginning to feel. He felt lost, not knowing what he really wanted to do with his life. He was searching for some meaning.

Last spring when Tom received the Bonfire Lesson invitation from Uncle Jack he placed it in a pile of other mail in his kitchen. At the time he did not give the letter much thought. However, what he didn't realize at that moment was that he was fooling himself, thinking that life was going just as it should. That his struggles in life would turn-around and that satisfaction was just around the corner. But in the next few weeks as the daily grind wore him down to reality, he began to open his mind to the possibility of taking the whole summer off and spending that first week in Pine Ridge with Uncle Jack. He loved Uncle Jack and he knew that a dose of his medicine was just what he needed to get his life back on track.

The freedom he experienced while driving to Pine Ridge reminded him of the feeling he had when school was let out for the summer and he was free from the confines of his class-room. As the car rounded the last bend before the cottage, his heart skipped a beat with excitement. He was reliving the hundreds of trips to the cottage with his family; gazing out the window at the sights and sounds of the place he loved the most. He pulled into the driveway, which was somewhat over-grown with grass and brush, and pulled up to the cot-tage. He was looking forward to relaxing, re-thinking his life and spending some time talking to Uncle Jack. Tom stepped out of the car and took a deep breath of the Pine Ridge air, drinking in the familiar smell of the pine forest. The chip-munks and birds flitted about almost as if they were excited to see him.

Tom left his luggage in the car and immediately headed down to the river to reconnect with the surroundings. He quickly followed the path to the top of the cliff in front of his parent's cottage and then went down the steps to the beach. As he walked along the deserted beach he passed the fire pit where many summer evenings were spent singing songs, toasting marshmallows, swapping stories and listening to Uncle Jack. Tom tossed down his t-shirt, sunglasses, wallet and removed his sandals, and waded into the water letting all the stresses in his life wash down the river. As he floated on his back he looked up at the majestic pines on top of the cliff and watched as an eagle soared effortlessly overhead. A euphoric feeling engulfed him. This was the place he loved the most.

That same day in another city, college exams were completed and the summer vacation was in full swing. The van was packed and the family was headed for Pine Ridge. Brian and Carolyn were full of excitement as they sat in the back of the van as they did on other summer vacations in their past. Back when they were pre-teens, the trip seemed far too long, as they were crammed with their parents and their summer belongings in their family vehicle. However, now that they were away from the pressures of college exams and their summer jobs, the drive was a bonus. It was a chance to read and to catch up on their sleep and they found that they actually enjoyed chatting with their parents from time to time. As the trip progressed a serene quietness entered the vehicle as both Brian and Carolyn slowly slid into their own thoughts, oblivious to the landscape surrounding them, remembering all of the pleasures of the past summers at the cottage and generating new ideas for their upcoming summer break.

Brian is a very spirited 21 year-old college senior who loves to experience life by doing it all. He's about 6 foot 2 with sandy brown hair and a slim build. Carolyn is a 19 year-old college sophomore with long blonde hair and an athletic build from years of dance. She cannot seem to find enough hours in the day to engage in some type of activity with her friends.

Long after the sun had disappeared over the horizon the family van approached the final road into the cottage. Despite the long hypnotic drive Brian and Carolyn were now at full attention peering out of the windows to see who's cottage lights were on indicating which families had arrived earlier. This was a ritual of their cottage trips. They were looking forward to catching up with their old summer vacation friends swapping stories of the past few years at college.

Brian thought back to the day he received the invitation from Uncle Jack. It was very timely. He had been struggling with the thought that next year he would be entering the work world. He was confused about this next big step in his life. He wasn't sure what he wanted to do. He was conflicted with many thoughts. He couldn't get excited about anything specific that he was studying in school. Living independently, money worries and his personal dreams revolved in his head. His dreams were not about conventional business like many of his friends, they were about spending his time in snow parks snowboarding. This was his passion. His dreams were about jumping, rotating and grinding rails and becoming a world recognized champion.

Despite being a sophomore in college Carolyn was already struggling with similar issues. Her love of dance had driven her to her fine arts college program. Yet she wondered how she was going to distinguish herself enough to graduate at the top of her class and how she was going to establish a career in dance. She had no model to follow. Both her parents took on more traditional professions and were at a loss to provide appropriate guidance.

It had been during spring break when they received Uncle Jack's invitation to Bonfire Lessons. It was also at this time that they had some heart to heart discussions with their parents and decided that they would ask for a week off from their summer jobs and join their parents at the cottage for a week. Somehow they knew intuitively it would be a life altering experience.

Uncle Jack was a friend and a mentor to both while they were growing up. He had always been there for them providing a safe environment to ask questions, listening and offering heartfelt advice. He had this uncanny ability to make them see things more clearly. Not only was Uncle Jack a mentor to Carolyn and Brian, he also was a guide for their parents who were graduates of Bonfire Lessons. Their parents were thrilled that they were going to join them at the cottage for the week, as they too sensed that a week with Uncle Jack was exactly what their children needed to set them on the right path.

As they rounded the last bend in the long gravel potholed road in the majestic pine forest, the last two residences had lights blaring like welcoming beacons for the traveling family. The last dwelling was their destination. As they passed Uncle Jack's home only a few feet from their grandparent's cottage the kids felt a rush throughout their bodies. They were at the cottage. As the van made its final turn in the driveway, the lights pierced through the kitchen window and within seconds Granny's head was visible, with a wide smile and hearty wave to greet them. Pine Ridge was her home for the summer and the travelers were excited to see her.

It was a joyful reunion filled with hugs and kisses at the opened back hatch of the van. In an excited, if not somewhat haphazard fashion, luggage and food were shuttled into the cottage. There was only half an hour of energy left in the travelers and everyone talked at once trying to catch up on the important current events. Before they knew it the clock on the wall chimed 11 p.m., time for everyone to retire for the evening, but just before sliding into their cozy beds the family stood on the deck to take one last look at the magnificent universe that unveiled itself before them. They stood together in silence, which was punctuated by the rhythmic serenade of the crickets. The cool evening breeze brought the mild fragrance of the pine forest to the forefront of their senses. The sights, the feelings, and now the smells confirmed the energy they felt. Their arrival had played out exactly as they thought it would.

The first day went by quickly as they experienced all that

Pine Ridge and its inhabitants had to offer. All of the pent up wishes over the year were acted upon. With the summer vacation in full swing the rafts were no longer beckoning the children, they were dancing in the waves as the children jumped and dove into the water playing tag and 'king of the raft' for hours. The fishing rods were active along the river as their bait was being cast into the river along its edge. And of course, as dusk appeared Uncle Jack lit the nightly bonfire as the brilliant sunset cast its remaining hues.

The smoke from the bonfire was a calling for most to end their daytime activities and for the kids and others to gather round; to sing songs, enjoy the warmth as the cooler evening set in, to laugh about the day and the collective history of the events at Pine Ridge. Uncle Jack led the group, as was now a Pine Ridge custom, in a medley of songs, many of which went back to the 1800s, however, the younger folk made sure that they mixed in some recent music just to balance things out.

As the evening wore on most people returned to their cottages and all that were left on the beach were Tom, Brian, Carolyn and Uncle Jack. It was the first opportunity to collect in a small group with Uncle Jack. They each sat around the fire staring into the flames with the sense that the moment would last forever; it was one of those golden moments of summer.

Key messages or ideas to take away from this Chapter:
A mentor is defined by Webster's dictionary as a wise or faithful adviser. Uncle Jack is a mentor to his students.

1. Who do you trust, respect and view as a mentor? Is it a family member, friend, or work associate?

2. When will you meet with your mentor next to discuss real issues and dreams in your life?

3. If you have never developed a mentor, where would you develop one? Who would you ask?

4. Who looks to you as a mentor in their life? Providing a safe trusting relationship where they can confide in you.

5. When, or to whom will you provide this gift of time, energy and love through mentorship?

6. Where is that special place that you can go to get a different perspective on life? Is it a park, campsite, beach, mountain cabin, cottage, river, lake, ocean?

7. When will you commit to going there to spend time in reflective thought? If something is holding you back, what is it?

2 *The Big Bang*

> Most of the fundamental ideas of science are
> essentially simple and may as a rule be expressed in
> a language comprehensible to everyone.
> —Albert Einstein

Uncle Jack broke the silence by formally greeting his invitees to the Bonfire Lesson with big bear hugs and kisses. The group caught up with each other's lives and shared their joy of being at Pine Ridge. There was excitement in the air as Tom, Brian and Carolyn were about to start what they considered a very important event in their lives. Many of the people they respected the most, including their parents, had attended Uncle Jack's Bonfire Lessons. Everyone spoke highly of the week and its impact on his or her life. It was considered a rite of passage for those that lived in Pine Ridge and for many others who had travelled there from around the world. With this knowledge they felt humbled to have this opportunity.

Interestingly, Uncle Jack has a keen sense of when to invite people to his Bonfire Lessons. There doesn't seem to be a defined age that generates an invitation; it seems to come to people when they need it the most.

Eventually the group found their own space around the fire and lay on their backs with their hands clasped behind their heads gazing at the stars. They started to relax, enjoying the moment, but also trying to understand the significance of Pine Ridge and all its beauty in context with the universe above them. As the bonfire became quieter turning to red and orange glowing coals, with only a few flames, the brilliance overhead seemed to intensify. Uncle Jack recognized that this energetic, constantly moving group was now entering a trance like state and as it always happened, the quiet would break with some thought-provoking question. Uncle Jack knew that when you reach this quiet state you connect with your sub-conscious mind and your higher self and that's when the big thoughts and ideas come to your conscious mind. Earlier Uncle Jack had introduced the evening's theme: "the universe." They were all now contemplating this thought in silence.

Eventually Brian broke the trance, "Uncle Jack, what's out there past the last stars we see on the other side of the universe?"

"That's a great question Brian and it is a great start for our discussion tonight. For tonight I want us to explore the universe we live in and to resolve some unanswered questions that you may have. It is my belief that to go forward in our lives we must understand the context of the magical universe in which we live," said Uncle Jack.

He continued, "I would like to start with some basic building blocks, such as the fact that everything in our universe is made of the same stuff. This is a very important concept that we will touch upon the entire week."

Carolyn asked, "Everything?"

"Yes," Uncle Jack responded. And then he continued, "not only are the sun, stars and the moon made of the same stuff but so is everything else in the universe. In fact, you are made of the same elements that make up the planets and the stars. Everything is made up of the same basic elements such as: carbon, hydrogen, oxygen, etc. This means your body, the beach we are lying on, the trees, the grass, the water, the animals, the things we manufacture like food, computers, as well as the stars, the sun, the moon, comets and asteroids. Everything is made from the same stuff."

"How is that possible?" asked Carolyn.

"There have been many observations and theories that have been generated by scientists and physicists over time that have led us to this conclusion. Have you ever heard of the Big Bang theory?" asked Uncle Jack.

The group responded with a nod of their heads.

He went on, "it is theorized that the Big Bang was the moment that the universe was created. It was created from a speck smaller than a period at the end of a sentence in a book. It is the source of the entire universe. This speck is called a singularity in the physics world. Based on the evidence found within our universe, many of the world's leading physicists have established that the singularity is the genesis of our universe. Everything that is in the universe, as we understand it, was contained in this dense speck. I know it's very hard to understand this, but bear with me for a moment. The effects of the event that we call the Big Bang are still occurring. The entire universe is still expanding. The Big Bang set everything in motion, such that all objects are moving further away from that starting point. You can think of it as an explosion wherein the effects go on forever. In the 1970's Stephen Hawking and Roger Penrose utilized Albert Einstein's General Theory of Relativity as a basis to prove that the universe started from a singularity. We will talk about Einstein and his contribution to science in great detail another night."

"Of course we've all heard of Einstein. Isn't Stephen Hawking the physicist who uses a computer to help him speak because of some debilitating illness?" asked Brian.

"You are right. Let me give you a quick biography of these three men and then I'll get back to the expanding universe."

"Albert Einstein is probably the world's most well recognized physicist. He was born in 1879 in Germany and died in 1955 in Princeton New Jersey where he spent the last few decades of his life. He truly was an alternative thinker who was able to synthesize much of the work done by others before him into his famous equation $E=mc^2$ and his two major theories of relativity: The Special Theory of Relativity (1905) and the General Theory of Relativity (1915). Again, we will be talking more about his theories over the next few evenings."

"Stephen Hawking was born in England in 1942. He studied the General Theory of Relativity and Cosmology at Cambridge University in the early 1960's. In 1963, he was stricken with Lou Gehrig's disease, but he defied all odds early in his illness and completed his education in 1966. He focused much of his attention in the areas of: singularities, black holes and quantum mechanics. Stephen Hawking eventually published the biggest selling physics book for the mass audience called A Brief History of Time in 1988. And as Brian mentioned, he is confined to a wheelchair and must use a special computer to synthesize his voice. In fact, before his famous book was finally published, he had to use the voice synthesizer to help him complete the final draft."

"Roger Penrose, was born in England in 1931. Penrose's specialty was mathematics. He and Stephen Hawkings worked collaboratively on projects related to cosmology. Penrose also worked on theories to unite relativity and quantum mechanics. He is the author of a number of books."

"Let me get back to our subject at hand – the expanding universe and how Hawking and Penrose helped us understand that."

"Thanks for the background," said Brian.

"There were many people who hypothesized that the universe was conceived from a singularity but it was Penrose who looked at it from different perspective. He studied how stars become singularities or what we now call black holes. As stars age they undergo tremendous changes in their lifetime, they eventually collapse upon themselves due to extreme gravity. He called this gravitational collapse."[1]

"Gravitational collapse?" questioned Brian.

Uncle Jack continued, "Yes gravitational collapse of a star occurs as the star gets older and its fuel runs out. The fuel for a star is helium, not petroleum gas that we use for our cars. A star is like a controlled hydrogen bomb explosion, where the hydrogen atoms collide and coalesce to form helium. It is the heat created in this reaction that makes the star shine. Eventually a star runs out of hydrogen and other nuclear fuels, it starts to cool and then contract. It actually goes through a series of contractions and expansions before its final death."[2]

"Penrose was able to prove that a star could collapse onto itself such that all its matter would be compressed into a singularity, such that the density of matter and the curvature of space-time becomes infinite – in essence a black hole.[3] You can visualize this when you think of a hot air balloon. When the heater is turned on the balloon is filled with air. When the heater is turned off the hot air balloon collapses. The difference with a star is that it eventually disappears into an infinitely small and very dense spec. Only very large stars can collapse into a black hole. Stars like our sun are not large enough to do so. They will collapse into dwarf stars or neutron stars, which are very bright stars burning their remaining fuel in the night sky."[4]

"Steven Hawking turned Penrose's theory around completely 180 degrees and believed that you could reverse that process much like watching a movie backwards on your DVD player. By reversing the process of a star's collapse into a black hole, it became an expansion from a singularity. Therefore, Hawking theorized that an expanding universe must have begun with a singularity."[5]

"We know that black holes exist in our universe because astronomers can see unique relationships between heavenly bodies. Objects appear to be orbiting regions of space as if a massive star existed but none is visible. The only explanation for this appearance is that a black hole is emitting its tremendous gravitational influence on the surrounding visible matter."[6]

Uncle Jack continued, "Since we know that eventually stars run out of fuel there are many black holes yet undiscovered in our known universe. Stephen Hawking theorizes that there may be more black holes in the universe than there are stars." [7]

"It's incredible to think that there are more black holes than stars," said Brian.

"I agree," said Uncle Jack. "Although it is not entirely surprising, considering where we are as a society in understanding our universal surroundings. Since we can't see black holes, we can only assume their presence by observations. Of course in the early days of star gazing, if it wasn't there to be seen, it didn't exist as far as anyone in the science world was concerned."

"Of course, there were many other clues to help Penrose and Hawking reach the conclusion that the universe was expanding. Outside of Albert Einstein's General Theory of Relativity, they used some of the early work from the likes of Edwin Hubble. By the way, you may have heard of the name Hubble before as the scientific community named the Hubble telescope after him. It is the Hubble telescope which orbits the earth and peers out to the stars. Hubble was an American astronomer who lived from 1889 to 1953."

"During his observations of the universe in the 1920's, Edwin Hubble was able to conclude that the universe is expanding.[8] He published this in his findings in 1929. Hubble was able to reach his conclusion by peering into the night sky and observing the stars in galaxies many light years away."

"How did he do that?" asked Carolyn.

"Hubble utilized some basic principles of light. I am sure you have looked through prisms in your elementary school days and what you would have found is that light consists of different wavelengths that appear as different colors to the naked eye. We can prove this when we reflect light through a prism. This same effect is found when we see a rainbow in the sky, as the sun shines through water droplets in the sky refracting the light into its different wavelengths. What Hubble knew from earlier scientists was that when you observe a light source such as car headlights moving towards you, the wavelengths of light are shorter and their spectra will be shifted to the blue end of the spectrum. When you observe car lights moving away from you, the wavelengths are longer and their spectra shifts to the red end of the spectrum. Therefore, when Hubble observed red shifts in other galaxies he concluded that these distant galaxies were actually moving away from us. In fact, what he found was that the more distant from earth the galaxy was, the faster it was moving away from us, just as Einstein's theory of relativity predicted."[9]

"The relationship between wavelength and speed is called the Doppler effect. You can also experience this effect with sound waves. When a motorboat approaches it has a high-pitched sound, which are shorter wavelengths of sound. As it moves away from you the sounds are much lower in pitch, which are longer wavelengths of sound. Listen for this the next time you are on the beach and a boat passes or even when a car or motorcycle passes by on the road."

"In conclusion, Hubble's observations suggested that at some point everything in the universe was close together. In fact it appears that 12-14 billion years ago that the universe was infinitesimally small and infinitely dense — a singularity."[10,11]

"If it is still happening, are you saying that we are in the middle of an explosion?" Carolyn asked.

Uncle Jack responded, "I was just using that as a way to describe how the initial Big Bang occurred and how things move out from the center of an explosion. But, in a sense that is the case. We are at a different stage, of course. As you can imagine, when this Big Bang occurred everything was extremely hot. There were no atoms in the early universe, only free electrons and nuclei. As the universe cooled by a few thousand degrees, protons and electrons combined to form neutral hydrogen. This was thought to have occurred 400,000 years after the Big Bang. The universe was about 1/1100 of its current size. In fact, cosmologists can measure cosmic microwave background radiation, which is the remnant of the heat left over from the Big Bang. It is uniform in every direction in the universe. Again, this remnant radiation is as predicted by physicists that theorized the Big Bang."[12]

"So, we've discussed singularities, black holes and the Big Bang let's discuss how stars like our sun are formed. Stars are actually hatched from the billions of galaxies that stretch across the universe. In the early days of the universe, it was expanding and cooling. Eventually some areas became denser than others as everything wasn't distributed evenly from the initial Big Bang. In some areas, the matter would start to collapse and the gravitational forces from outside these areas would start them to rotate, eventually forming into disk-like galaxies, which are what we see in the universe today. Stars eventually formed within these galaxies when large amounts of hydrogen and helium gas started to collapse in on themselves due to gravitational attraction. As the atoms of the gas collided they would continue to do so increasingly until they would become so hot that nuclear fusion would occur. As more hydrogen fueled this nuclear reaction, pressure would build up to a state such that a stable star would be created. However, as the star's fuel diminished the star would contract and it would convert the lighter helium into heavier elements like carbon or oxygen.[13] Massive stars may take 100 million years for the nuclear fuel to run out. It is predicted that our sun may last another 5 billion years. Interestingly enough, smaller stars outlive the larger ones."[14]

"I guess we don't have to be worried about being frozen in the dark for a few billion years, as if that length of time matters much to me," said Carolyn.

"You are right Carolyn, the lengths of time during which major events take place in the universe make our own lifetimes seem like a short breath of air," said Uncle Jack.

Uncle Jack continued. "As stars age, they cool down further. They eventually collapse into a very dense state such as a neutron star or dwarf star (if it's not a massive star) or if they are larger stars they transform into black holes. Sometimes during the life of a star some outer regions would get blown off in a tremendous explosion called supernovas. The heavier elements would be thrust out into the universe and become raw material to form other stars or planets such as those that orbit the sun.[15] The planets that revolve around our sun are made up of material spewed out of the sun itself. Remember that the earth was a hot molten mass that has cooled down over time. We are reminded about this when we read of a volcanic eruption occurring somewhere on our planet. The molten magma stored in the core of our planet spews out onto the surface of our planet, creating new land like the Hawaiian Islands. We are also reminded of this when there are earthquakes caused by tectonic plate activity in which massive pieces or plates on our earth shift on this molten mass."

"Wow, I had no idea that the planets were made from the stars. Now I understand what you mean by the fact that we are all made of star stuff," exclaimed Carolyn.

Brian added, "I never really understood the development of our universe either, it really is a fascinating story Uncle Jack...please keep going."

Uncle Jack continued. "In our solar system, all the planets and other space debris like asteroids and dust, orbit the sun, which is a star. However, the sun is also in a position relative to other major heavenly bodies, which are held in position by gravitational forces that we do not fully understand at this time, but clearly the nuclear disk at the center of our galaxy, the Milky Way, affects it from a gravitational perspective. All of the other galaxies in our known universe affect each other to a lesser degree. At this point in time, with the aid of the Hubble Telescope it is thought that there are at least 50 billion galaxies in the known universe. There have been some estimates that suggest that the number of galaxies may be well over 100 billion galaxies. Furthermore our own galaxy, the Milky Way, is estimated to have approximately 100 billion stars."

"Wow," said Brian. "I had no idea that there were 100 billion stars in our own galaxy and that there were 50 billion galaxies. Just think about how many planets there are?"

Uncle Jack continued, "These galaxies are not random but are found in clusters throughout. We now call them superclusters. The Milky Way galaxy is a member of the Virgo galaxy supercluster, which is made up of approximately 5,000 other galaxies much like the Milky Way. The Virgo galaxy supercluster is about 100 million light years across.[16] This is massive. In order for you to get a sense of this distance, understand that it would take 100 million years for light to travel from one side of the Virgo galaxy supercluster to the other."

"Here is another way of getting some sense of the vast distance and the incredible speed of light. The distance to our sun is 8 light minutes. Eight minutes is the length of time it takes for light to travel from the sun to the earth. In fact, in one second light travels 186,000 miles or 300,000 kilometers.[17] In one light year, light would travel 194.89 billion miles."

"Uncle Jack these distances are mind boggling but that example really gives me a sense of scale. Eight minutes for light to reach the earth from the sun and 100 million years for light to travel across the Virgo galaxy supercluster? It's hard to fathom distances that great," said Brian.

"As you look up to the sky above us, you see the stars in the milky way. Some of the stars we see no longer exist. The light that we see from some of these stars existed millions or billions of years ago, as our home galaxy is approximately 13 billion years old. The closest star to the sun is Alpha Centauri, which is 4.3 light years away. The closest galaxy to the milky way is the Andromeda Galaxy."[18]

"Let's talk a bit more about our home galaxy, the Milky Way, to give you some sense of the size. All stars in the Milky Way revolve around a nuclear region or nuclear bulge, which is at our galaxy's center. It is theorized that at the center of our galaxy and every other spiral galaxy is a giant black hole. It is also theorized that in the nuclear bulge approximately 10 to 1000 stars occupy the same space as our sun does. Much like the colliding atoms in the sun, it is thought that there are many near collisions. Our sun does not have this same problem. It sits about two thirds of the way from the nuclear bulge in the Milky Way's center. In the last billion years no other star has come within one quarter of a light year of it.[19] Imagine the 100 billion stars in the Milky Way all revolving like planets revolve around the sun."

"The nuclear bulge of the Milky Way is the equivalent to our sun in the center of our universe of planets. The remainder of the stars in the Milky Way are positioned in a pinwheel fashion, with blades of stars, including our sun on one of those blades, two thirds of the way out from the nuclear bulge."[20]

"Wow, Uncle Jack the space and dimension of the known universe is mind-boggling. It's hard enough to comprehend hundreds of miles between destinations as we travel in a car, let alone the distances between stars and galaxies. There are such vast empty spaces and it is hard to link everything up there together, because we are so small," said Brian, in amazement again.

2.1 *Our Ideas are Challenged*

The joy in looking and comprehending is nature's
most beautiful gift.

—Albert Einstein

"Another thought to consider is that most of the universe, or
what we sometimes call space, is a void."

"What do you mean by a void, Uncle Jack?" asked Carolyn.

"A void essentially means nothing, or absence of something. Between the earth and the moon there is nothing other than sporadic space debris like dust, rocks and energy,
which is not visible to the naked eye. Do you remember
when I said everything is made up of the same stuff, essentially star stuff?" They nodded. "Well, we think of space
like the air that surrounds us on earth, however, while it
looks like there is nothing in the air that we breathe, it is in
fact made up of gases that have the same elements that are
part of the make-up of the universe, such as oxygen, carbon
dioxide, and hydrogen. On earth there are approximately10
million trillion atoms in a cubic centimeter of air at sea level.
Between the objects in space there isn't much of anything.
It is essentially a vacuum, void of almost anything. Between
the stars in our universe there is less than one atom for every cubic centimeter. The space between galaxies is about
one atom for every cubic meter.[21] So its not a true vacuum,
but very close. However, there are other energies at work
in this void, like gravity, and of course light and energy and

objects pass through it."

"It is currently theorized by many leading physicists that the universe is actually filled with dark energy. This dark energy is approximately three quarters of all of the mass in the universe. It is energy that is best described as anti-gravity. Dark energy opposes gravitational effects and permits the expansion of the universe. Unfortunately, like the theoretical graviton as proposed by Einstein, we are unable to physically prove its material existence. This continues to be one of the greatest mysteries of the universe."

"So, from a simplistic perspective, what is gravity? I'll ask that question for you because I think it needs some explaining. At a basic level, very large objects attract smaller objects. In fact, as we all know, gravity is the force that holds us all on the earth's surface. Any object on the planet tends to fall to the earth if there isn't something holding it away from the ground. Back in the 17th century, Sir Isaac Newton first described gravity by using the analogy of an apple tree. Once the stem is no longer able to support the weight of an apple, it is gravity that forces the apple to fall to the ground. If it is true that large objects attract smaller objects, you would think that the sun, being much bigger than the earth, would force the apple to go towards the sun. Of course, we know that this is not the case in reality. Why do you think this is so?"

"Because apples don't fly Uncle Jack," quipped Brian.

"Very funny Brian, you are the original good humor man. We always have a good time when we have these bonfire discussions." Uncle Jack continued, "Great distances weaken the effects of gravity. Even though the sun's gravitational force keeps earth in its orbit, everything on earth is affected by the local gravitational field. And as we all know, the moon orbits the earth demonstrating its powerful gravitational force despite the fact that it is approximately 350,000 kilometers away."

"Around any heavenly body like stars, planets or moons there are five locations called Lagrange points, which are named after the French mathematician who discovered this fact. These points are places where one heavenly body's gravity balances against another heavenly body. As an example, spacecraft can move between a planet's gravitational pull and the sun's gravitational pull without using a lot of fuel because the craft does not have to combat the effects of gravity pulling it in either direction."

"Water as we find in the river, gives us great evidence of the effects of gravity, as it always finds the lowest point on the earth. The river water in front of us will eventually empty into the ocean, which is the lowest point of elevation on the planet with a few minor exceptions."

"Yes, like the Dead Sea and Death Valley desert in California for example," said Brian, pleased to be able to add to Uncle Jack's lesson.

"Correct Brian, now, how do you combat gravity's effects? And, what is an orbit? Let me get into this quickly before the 'good humor man' has a chance to start up again."

"Humans would remain like babies crawling or rolling on the ground, if we did not grow and develop our leg muscles, which provide us with the strength to stand up and walk. We are still glued to the ground by our feet but we are able to pick up the rest of our body. It takes a great deal of work to pick up our bodies and walk, run and climb stairs. We all know how much energy it takes to run up stairs or to climb the cliffs beside the river. That is because we are going against gravity. That's why we need to maintain our fitness levels to be able to combat gravity. Then, we need to replenish that energy with fuel, which we call food. Have you ever noticed, how it takes very little effort to jump from the cliff into the river? That's because we are working *with* gravity, not *against* it. All you need to do is jump.

Carolyn exclaimed, "I don't find it easy to jump off the cliff into the river."

Everyone laughed, including Uncle Jack. "Now, there may be other energies that come into play to make it difficult to make that jump, but we'll get into that some other night.

That's the interesting stuff."

"Now back to my definition of gravity. Large objects do attract smaller objects, but that doesn't mean they always crash together like an apple falling off the tree to the surface of the earth. Large objects do attract smaller objects but there is a balance point. Other forces such as centrifugal force tend to push an orbiting object away from the gravitational pull of the larger one. That's why the earth doesn't crash into the sun or the moon doesn't crash into earth. They are orbiting the larger heavenly bodies. One force pulls it to the larger object and another force is trying to make it escape from its orbit."

"That's how we are able to put communication satellites or the international space station into orbit around the earth without them crashing back to earth. Once they are in the outer atmosphere at a certain distance they orbit the earth or remain stationary depending on the distance. They will stay there much like the moon orbits the earth. However, occasionally satellites get into unstable orbits and they begin to become prey for the earth's gravity. If they cannot be controlled remotely from tracking stations on the earth they will come crashing down somewhere on the earth's surface."

"I have seen documentaries on television explaining how they have tracked incoming rogue satellites or the early space station called Skylab. They say that these are dangerous events when they come crashing to the earth, but there is so much unoccupied land on earth it's highly unlikely that it would hit anything," said Brian.

"Thanks for that example Brian. Gravity was thoroughly explained by Issac Newton, however, there is a more advanced perspective called space-time, which is described by Einstein's General Theory of relativity that is a more accurate reflection of what gravity intends to describe. We will save that for another night. We owe much debt to the insightful knowledge of Issac Newton who lived in England, from 1643 to 1727. Newton spent most of his adult life putting sense to

the material world we live in. In 1687 Newton published a
book called 'Pyhlosophiae naturalis principia mathematica',
arguably one of the greatest scientific books every published.
His work and his life focused on describing what we call
the Newtonian world. It is the material world we feel, see
and touch. He taught us about gravity and how the planets
were attracted to the sun. He really had some ideas that were
ground breaking at that time in history."

"Earlier you asked the question — what is beyond our
known universe? Well, let's continue to understand what's in
it and what we know about it before I answer that question,"
said Uncle Jack. He continued, "Current estimates suggest
that the edge of the universe is about 15 billion light years
away. This conclusion is based on our assumption that the
Universe is approximately 15 billion years old from its birth
from the singularity we described earlier.[22] It is thought that
the Milky Way galaxy is approximately 13 billion years old
and, as I mentioned earlier, the sun is about 5 billion years
old. As the astronomers peer at the heavens, they are viewing
ancient history. The further they look into space the further
back in time they see. It is like a time machine of sorts. The
light that we see from stars that are 10 billion light years away
is the light created shortly after they were born. The sun did
not exist at that time, and our own galaxy was just forming."

"To get an idea of the youth of our galaxy and our sun and
solar system you need to consider the following. It takes 24
hours for the earth to go through one revolution, then the
earth takes 365 days to revolve around the sun but it takes
the sun 220 million years to revolve around the nucleus of
the Milky Way. In fact, our sun has made fewer than 25 of
these trips.[23] When instruments probe deep space, into the 15
billion light-year range they record nothing but a dull energy
haze called the microwave background. This is thought to be
the remnant glow of the creation explosion."[24]

"So what is beyond the edge? It is impossible to tell because that is the beginning of time as we know it. Anything

before that goes back to the singularity itself and that cannot be explained or fathomed. This becomes the realm of God and we don't yet have the knowledge to explain it. To ask the question; *what is beyond the universe*, is synonymous with the question; *what was there before the singularity?*' Brian, I can only tell you what I know and what we hypothesize, however, what I do know with much certainty is that there is a lot more for us to learn about life and that this learning will go on infinitely."

"You're not kidding," chuckled Brian.

"So the sequence follows that the singularity hatched the galaxies, that hatched the stars, that hatched the planets, that eventually led to the evolution of life as we know it. Thus began what we call living things, from microscopic life, to plants, animals and, of course, human beings."

"But what was there before? This is a question that perplexes us. This is a question that we cannot answer. Especially when you consider the information that our scientists and physicists have pieced together. When you start to question the boundaries of our universe: either beyond it or before it was born, you are entering a realm of understanding that we can't conceptualize. One of the theories is that we are in a repeating cycle wherein the big bang has erupted infinite times and within each cycle is the big crunch. If you could envision, for a moment, black holes forming throughout the universe that begin to munch up all matter everywhere, much like the Pac Man video game. Everything is pulled into their tremendous gravitational fields until larger and larger black holes eventually consume everything until we are back to a singularity.[25] We cannot prove that we ever had these repeating cycles because we cannot get any information beyond the Big Bang. So, the theory remains a possibility, waiting for further evidence to sway it one way or another."

"Hey Uncle Jack, even if we were in a repeating cycle, shouldn't there be something there to create it in the first place?" asked Carolyn.

"Well Carolyn, again we can't answer this question and it

is at this point that it moves into the realm of God and for those of us on earth it helps us believe in magic. This is why we must accept magic in our lives. Like the ability to live the life we wish, to strive for worthy goals that make a difference to ourselves and those we love. The magic becomes apparent when we explore the question about what exists beyond the boundaries of the *known universe*. The key to this question is that it focuses on the *known universe*. What I mean is that it focuses on only what we know, not on what we are not yet aware. What if our universe is one of billions of others? Perhaps other universes are being hatched from singularities throughout an infinitely larger universe, each one at a different stage or a similar stage to ours?"

"Throughout our lives we are conditioned to experience a beginning and then an end, whether that is reading a book, building a house, or the evolution of life on our planet. We can understand when something has an inception point. *What was there before the singularity?* In our human experience there always is some trigger for everything we experience. Who and what made the singularity erupt? What was the trigger and where was the singularity before the Big Bang? What holds the known universe?"

"Currently we can only peer to its edge, which is peering into the past, to the time of the formation of our universe. Even if we consider the suggestion that our known universe is one of an infinite number of universes expanding in infinite space, the question always comes back to: Where are the edges? What's on the other side? Who started this whole process and how?"

"Again, we simply can't conceptualize the answers to these questions. If God created the universe, who was there before God? Is there a God? Where did God come from? We have been taught that God is everlasting in all religions. How is that so?"

"All of these answers are impossible to conceptualize with our common human toolset. Therefore, we must believe that life is magical because it is beyond our understanding. That is the basis of most religions. God or as some refer, the Supreme Being, is the creator of the universe and that is beyond our human comprehension."

The group was enraptured, listening, and pondering but not able to answer the questions. Uncle Jack continued.

"Perhaps even hearing this will make you a little uneasy because your mind cannot conceptualize it and in our humanness we need to think of things as having an origin. But all of this continues to beg the question; what was there before? If you can get this subtle point, or admit to yourself that we have a paradox, then magic is possible and we are living proof."

"At our collective level of thinking we are at the dawn of our knowledge. We are *pre-kindergarten* in our ideas. Our understanding of the spiritual world and the world of science are beginning to merge. We are much more than what we see, feel, and touch."

"The model of the universe has similar magical principles as found in the quantum mechanical world. We will learn more about quantum mechanics tomorrow, but in essence it focuses on the minute things that make up our material world, like electrons, bozons and quarks. Physicists are able to make mathematical models that explain our known universe, and others are able to do the same for quantum mechanical universe. But both models don't fit perfectly together into a unified model of life. This is something that physicists are exploring today. The most potent evidence of a unified theory is found in an evolving theory called String Theory, which attempts to unify the large and small theories of relativity and quantum mechanics respectively.[26] The quest for a unified model of life is an everlasting quest that we will endeavor to solve."

"It wasn't until the late 19th and 20th centuries that we began to go deeper into the unknown world, which we can't see or touch, by examining the work of many physicists and scientists, culminating with the Theory of Relativity, $E=mc^2$ and Quantum mechanics. What I call the world of magic."

As the fire turned into embers, Uncle Jack tells the students that he has talked enough for one evening. He closes by saying; "Tonight, Brian asked a simple question about the universe which took us down the path of our discussion. We are all made of star stuff, thanks to the Big Bang, and I have just covered the tip of the iceberg of what you need to know about our universe. There is so much more magic that occurs in our universe that we take for granted and if we are able to discover how this magic works we will be able to live the life of our dreams. So tonight as you sleep, your subconscious mind will absorb this evening's lesson in your dreams and prepare you for tomorrow night when we will get into the magical non-material world." And with that he said, "Good night" as he extinguished the fire with the water from the river.

Key messages or ideas to take away from this Chapter:

1. Everything in the universe, including humans, are made of the same ingredients such as carbon, hydrogen, and oxygen to name a few.

2. The universe is expanding and is approximately 15 billion years old. Our sun is approximately 5 billion years old.

3. At the edge of our universe (which is still expanding) there is a dull energy haze called the microwave background. It is the remnant glow of the creation explosion from a singularity, smaller than a spec at the end of this sentence.

4. There are approximately 50 billion galaxies in the known universe. In our home galaxy there are approximately 100 billion stars.

5. There are potentially more black holes than there are stars in the universe. Black holes are the remnants of very large stars, which have run out of fuel.

6. There are some significant questions that we cannot answer such as:
   § What is beyond the edge of our universe – on the other side?
   § If the universe started as a singularity, what was there before it; what initiated the big bang?

7. There is much more to life than we know. If we can't conceptualize what existed before the singularity, then what created it? What is beyond the edge of the universe? These questions lead us to believe in *magic*, which is a spiritual dimension far beyond the Newtonian world we have grown up in.

3 *Quantum Mechanics*

If quantum mechanics hasn't shocked you, you haven't understood it yet.

—Neils Bohr

As the sun slipped away for another day, the brilliant summer sky evolved into the beautiful red glow of the setting sun. As dusk turned to darkness, the first glimmer of stars appeared. Uncle Jack was building the campfire as the daylight activities came to an end and the group of kids old and young helped him gather wood for the fire, in certainty that once the coals were glowing their marshmallows would be ready for toasting. As the group collected driftwood on the beach, the discussion from the previous evening danced in their minds as they saw the universe unfold before their eyes. The vastness of the universe, the age and the distances to these remote stars was mind-boggling. Much of what they were seeing was how these stars looked billions of years ago because of the time it took for light to travel to the earth. Their lesson from the night before had obviously had a great impact.

Tom helped direct the children while Uncle Jack was stoking the fire. With darkness the campfire became the beacon that drew everyone together. Stories of the day were shared amongst those in attendance as well as perfectly toasted marshmallows. As always, there were friendly competitions to determine who could toast the best. The self-proclaimed marshmallow masters all explained to the young kids how to avoid burning their marshmallows by sticking it in the coals or having it spontaneously ignite like a fire-eaters torch. There were many laughs as marshmallows glowed as they ignited and turned to charcoal black before they could be blown out. The youngest of kids would always offer the grownups what they thought was a perfectly toasted marshmallow. No doubt they would be smeared with ashes or sprinkled with shale or sand from the beach. In a peculiar way this added to the fun of the experience. There was laughter and appreciation for each creation. As the treats disappeared the youngest children were taken up to their respective cottages and nestled into their beds to rest up for the next day. Meanwhile the regular crew of kids and adults lounged around the fire singing campfire songs. Eventually, it was only the students and Uncle Jack that remained.

As the fire reached its zenith they reflected again on their lesson of the previous evening. Tom and Carolyn were discussing how much energy it took to grow the trees that were used to fuel the fire which eventually were used to warm them in the cool summer night as well as to provide energy to toast their marshmallows. They related their lessons regarding the stars and the creation of the universe to the burning logs in the fire in front of them. The scale was completely different. The immensity of the stars was hard to fathom.

Uncle Jack waited until all that could be heard was the crackle of the fire, together with the crickets and frogs in the background occasionally serenading the group, before he spoke.

"We talked about the grandness and the immensity of the universe last night, tonight I want to focus on the opposite side of the spectrum. Last night I had mentioned the term quantum mechanics, it is the law that applies to the very small. When we look at the 'make up' of anything, for instance the wood in our fire or the clothes that we wear or even our bodies, scientists agree that these things are all made up of molecules and atoms that are made up of smaller elements such as carbon, hydrogen, helium etc. which are made up of even smaller particles like protons, neutrons and electrons. And then again, at an even smaller level there are bozons, quarks and leptons among other minute elementary particles."

Tom interjected, "I remember studying these elements in my high school chemistry class. We had to memorize the periodic tables that described all of the elements known at the time and how many electrons, protons and neutrons were in each. The only one I can remember is H_2O, which is water. There were two hydrogen atoms and one oxygen atom. I haven't much thought of that in the last 13 years, unless I was playing Trivial Pursuit."

"That's correct Tom," chuckled Uncle Jack. "There are 92 different elements that are known to humankind. Most people do study these elements in high school chemistry class, however, tonight I don't want to focus on the periodic tables and what you learned in high school chemistry class, I want to introduce you to the quantum world."

With that statement everyone around the fire knew they were in for something that was going to expand their thinking. It was becoming evident that this week was about much more than marshmallows and singing.

Brian broke the silence by asking, "What is the quantum world all about?"

"Well, as I said earlier, quantum mechanics describes the law of the minute scale. The behavior of these minute particles is very odd when compared to the very large objects that are under the influence of gravity. As time progressed, discoveries built upon each other to the point where Earnest Rutherford and Neils Bohr began to unravel the atom in the early 1900's. Scientists have long searched for, and continue to search for, the basic building blocks of matter. This search is to further understand the physical world from which everything emanates.

"Here's a bit of background on Bohr and Rutherford to give you a reference. Niels Bohr was born in Denmark 1885 and died in 1962. He worked with Earnest Rutherford after meeting him at Cambridge University in England. Rutherford was born in 1871 in New Zealand but spent most of his life in England where he died in 1937."

"Prior to the time of Rutherford and Bohr, there was no proof that there was anything smaller than an atom itself. Neils Bohr was the originator of the Bohr model, which illustrates electrons orbiting the nucleus of an atom much in the same way as we perceive the planets orbiting the sun.[1] However, at the quantum level gravity seems to have no effect as it does with the orbiting planets and galaxies. That is one of the distinct differences between the theory of relativity and quantum mechanics and the reason for the pursuit of a unified theory.[2] As we will soon learn Bohr's model wasn't completely correct. Bohr's initial work became a building block for scientists to discover the uniqueness of the quantum mechanical world."

"Rutherford deduced that the atom wasn't a solid object, as others had speculated in the past. He theorized that an atom was composed of mostly empty space, that there was a tiny speck of a nucleus at the center of the atom. This nucleus was far from the flurry of activity created by the electrons on the outer surface. He and his assistant James Chadwick determined that the nucleus not only contained protons but it also contained neutrons. To unleash the power of $E=mc^2$, Chadwick theorized that the key was to add more neutrons into the nucleus to destabilize the atom[3], but I will cover more about $E=mc^2$ on another night."

"Tonight I want to focus on quantum mechanics. The more that physicists conducted atomic experiments the more they realized that their basic concepts, their language and their whole way of thinking were inadequate to describe atomic phenomena."[4]

"The real birth of quantum mechanics happened in 1925 with Werner Heisenberg's initial work on Matrix mechanics, which eventually transformed into quantum mechanics.

Heisenberg was born in Germany in 1901 and worked most of his career there until his death in 1976.[5] In 1927 he produced a revolutionary theory called the Uncertainty Principle, which helped further develop his theory on quantum mechanics. In 1928 he finally published the book 'Physical Principles of Quantum Theory'. Heisenberg's discovery of quantum mechanics led him to receive the Nobel Prize for physics in 1932. The Uncertainty Principle in essence, describes how one can never be sure of both the position and velocity of a sub atomic particle. The more you know of one aspect the less accurate you can be of the other.[6] I realize that this may not seem revolutionary to you, but to the physicists at that time it was a critical new thought."

"So let me understand this point", interjected Tom. "The more you know about position, the less you know about the velocity and the more you know about the velocity the less you know about the position. Do I have that right?"

"You do indeed. Perhaps the easiest way to envision this might be to imagine if you were able to locate a particle. The very act of shining a light to see it more clearly, actually disturbs the velocity of the particle itself. This is the key to the uncertainty principle. I realize it may not be that obvious to you, but this was revolutionary for its time and still is today. Why? It removes the deterministic vision of the world.[7] Whereby you can never precisely measure the current state of the world. The mere act of observing it, changes it. I'd like to draw this into a real world example beyond the atomic level to our everyday lives. Imagine if one day a television producer asked you if you would like to participate in a program called, 'A Day in the Life'. In this program the television crew, with cameras in tow, would follow you for 24 hours to see what you did in your day. They would capture with whom you associated, when and where you ate and where you slept and many other intimate details. Do you think your day would be slightly different if you were observed? I don't think you could help it. This concept is essential for you to understand before we have our fireside chat about manifestation. Store that thought for then. This begins to paint the magic that gives you the ability to live the life you wish to live. By the way I'd like to point out that at this moment you are living the life you wish to live. Despite how you feel about this statement, it is true. You have manifested your world and we will discuss this further on another night."

All was quiet as they pondered this thought.

Before they could burrow deeper into that thought with questions, Uncle Jack jumped in.

"Let's get back to the basics of quantum mechanics again. The interesting thing about quantum mechanics is that even after Heisenberg developed his theories; its effect on the physicists' view of reality was truly shattering. Heisenberg asked himself if nature could be so absurd. The new physics necessitated profound changes in concepts of space, time, matter object and cause and effect; and because these concepts are so fundamental to our way of experiencing the world, their transformation came as a great shock."[8]

"As we get back to the search for the smallest building blocks that make up any material object we find that what we have learned through the laws of quantum mechanics is that there are no basic building blocks at all. I know this is probably a surprise to you; it was a huge shock to the physicists of the day. In the quantum world, relationship is the key determiner of everything. Subatomic particles come into form and are observed only as they are in relationship to something else. They do not exist as independent things."[9]

Tom interrupted, "I remember my physics teacher in high school talking about wacky things at the time. He was standing at the front of the classroom and he asked us what he was leaning on. We all yelled out funny answers, but truthfully he was leaning on a filing cabinet. So when someone shouted out 'filing cabinet' he said that it wasn't true. He said it was just our perception or our interpretation because there really wasn't a filing cabinet there. I didn't understand it at the time…because it sure looked like a filing cabinet to me."

They all laughed enjoying the break in what seemed like complex ideas.

"So I guess what you're saying Uncle Jack is that there are a bunch of reactions between things at the smallest level that only exist in relationship to each other to produce a material object?" Tom voiced this with uncertainty.

"You are right," said Uncle Jack. What you saw in that classroom was a filing cabinet and what we see in front of us tonight is the river, the rocks and sand on the beach, the cliffs, the trees, our fire and our friends. They are all real things, but they are made up of non-material objects that come in and out of existence continually. As Deepak Chopra describes it, your body is not separate from the body of the universe, because at the quantum mechanical levels there are no well-defined edges. If you had the ability to see differently, you might see that you are like a wiggle, a wave, a fluctuation, a convolution, a whirlpool, a localized disturbance in the larger quantum field. The larger quantum field is the universe to your extended body."[10]

"So each object is a localized field, which interacts with everything else, like the ecosystems you study in school and where we live. We are all made up of energy and information contained in the relationships between these subatomic particles. Looking at our life in a more holistic perspective and at how the ecosystem connects everything, it all begins to make much more sense. The connection to everything starts at the sub-atomic level. When it comes to our health, we can begin to see the links between the environment we live in and work in, and the people we associate with. They all affect us and we affect them, because we are all *connected*. I promise to speak more about how health can be compromised by what we think and do some other night but let's get back on track with our quantum physics shall we."

"Uncle Jack you are just teasing us with some of your conversations, I want to know more about how the environment fits in with quantum mechanics and health and performance," said Tom.

"Some people never understand this stuff and I want to make sure you see the big picture complete with all of the details. So for now you'll have to be patient," said Uncle Jack.

"It's just that it is all very exciting. There are so many thoughts and questions swirling around in my head."

"Great, that is what I'm expecting and I'm glad that you haven't disappointed me," replied Uncle Jack.

"Okay, let me continue with my earlier train of thought. All of the physicists and scientists of the day, including Einstein, Bohr, Rutherford, Heisenberg, James Chadwick and later many others like Stephen Hawking further developed their theories building on the incredible developments produced in the first 30 years of the 1900's. Incredible leaps in knowledge occurred during those early years of the 20th century through the special and general theories of relativity and quantum mechanics."

"These theories have revolutionized our lives, springing us forward past the Newtonian world that we thought we lived in. The challenge in life is that the general public doesn't understand the magic behind $E=mc^2$ and quantum mechanics. These ideas are locked in the minds of the great scientists and physicists of our day. Despite this fact, we currently use many of the products that leverage this knowledge. Science, technology, medicine and our perspective on spirituality has all benefited from these developments. The general population takes many things for granted and needs to take the time to contemplate the incredible potential provided in our universe to radically change their lives."

"We live in a paradox. On the one hand we experience our lives in a Newtonian world, one that was envisioned in the 1600's and yet we leverage products that were invented in the modern world of relativity and quantum mechanics. Having said that, we haven't fully taken advantage of the magic in our personal lives. The magic is there for all of us to use. I want to give you the key."

"I want to tell you about a really interesting experiment that tells an interesting tale about quantum mechanics. Really there are two experiments, one done in the 19th century and a second that has been duplicated many times in the 20th century. Thomas Young demonstrated the double-slit experiment early in the 19th century. His goal was to illustrate how light travels in waves.[11] His experiment requires two pieces of cardboard, a light and some type of screen or wall that can be projected onto. One piece of cardboard has one narrow slit while the other piece of cardboard has two narrow slits."

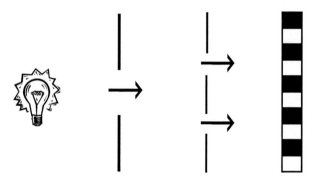

"If you shine the light through the first slit in the first cardboard sheet eventually light passes through the second sheet and illuminates a pattern of light and dark bands on the screen or wall. This is called an interference pattern. What this illustrates is that light travels in waves. The waves coming through the second piece of cardboard combine to generate the bright spots in the interference pattern and secondly some light waves interfere with each other and eventually cancel each other out leaving the dark spots. This is entirely explainable when we look at light in the form of waves.

This experiment is easy for anyone to experience. You can try it at home in a dark room with a flashlight and a couple pieces of cardboard paper with slits, as I described earlier."

"Now there is a second experiment that was done hundreds of years later using particles such as electrons and photons. Photons are the smallest unit of light."

"Stop for a moment and think about what would happen if we had a sheet of plywood down here on the beach and we cut a hole in it and then fired rocks at the hole, what would you expect to find?"

Tom answered, "You would get a pile of rocks on the other side of the hole."

"Correct," said Uncle Jack. "And if you had two holes, you would get...."

Carolyn answered, "Two piles of rocks behind both holes."

"Right again," Uncle Jack replied.

"So let's think of an electron as a rock. What do you think would happen if we fired individual electrons or photons in the double-slit experiment instead of light in waves?"

"I guess we would get a lot of light spots behind the first hole and after the second piece of cardboard..." Brian's response ended abruptly as he struggled to answer.

Uncle Jack jumped in, "If we used a screen, much like a television screen, to capture the electrons behind the two sheets of cardboard the result is that we have spots on the screen for every electron fired through. If you fire a few electrons you see random spots on the screen. As the experimenter continues to fire thousands or tens of thousands of electrons or photons more and more *apparent* random spots begin to appear on the detector screen. But something mysterious begins to happen. The screen begins to show an interference pattern as was found in the initial experiment with the light source that Thomas Young illustrated.[12] So what has happened here?"

The group was mesmerized, it was clear they didn't have an answer.

"Don't stop Uncle Jack, we know you know the answer. I don't think any of us is going to help you on this one," said Tom.

"Well, now you are experiencing exactly what the scientists experienced, which was complete confusion. It gets even more puzzling, but before I get there, let's discuss what we know about light. Is it a wave or is it a particle? This was one of the biggest puzzles of the early 1900's. Groups of scientists were out to prove that light was a wave, yet others set out to prove that light was a particle. Young's experiment demonstrated how light travels in waves. In the end, both groups were right. In Young's first experiment, we observed light behaving as a wave. The second set of slits produced two wave sources of light. In some cases they were in step, which created the bright spots, however, in others they cancelled each other out and did not illuminate the screen, which jointly created the interference pattern.[13] Are you with me so far?"

They nodded their heads, which was enough for Uncle Jack to continue. "In the second experiment, the electron started out as a particle, like the rocks we imagined throwing through the sheet of plywood. They ended up like a particle as they appeared as a spot on the screen. What happened in the middle? It was as if the electrons had some knowledge of the past and the future so that they could pass through the holes and land in the most appropriate spot to construct an interference pattern. It was as if the individual electrons or photons were all contributing to the pattern development together with some type of intelligence. And perhaps the electron or photon actually travels as a wave but begins and ends as a particle.[14] But how could they have this knowledge of the future which allowed them to work together to construct the interference pattern?"

"That is really weird," gasped Carolyn. And then they all chatted out loud echoing each other's questions and wondering how this could be possible and what the scientists had to say about this information.

"Well, as I mentioned, it does get weirder," said Uncle Jack. If he didn't have them all in his rapture before, they certainly were now. "The scientists, were also very puzzled themselves, so they tried to put detectors in place to find out what was happening. They wanted to find out which hole the particles were going through. As soon as they started to do that, the result of the experiment was two blobs of light on the screen. They were able to detect each particle and which hole it went through but the mysterious results of the interference pattern on the screen disappeared."[15]

"Okay, you are going to explain this so that it makes sense, right Uncle Jack?" asked Tom.

"Yes guys, I am going to try to explain it but it is still a mystery to the scientists as to why it happens. It is the law of quantum mechanics and that's that. It appears as if when we attempt to measure and detect exactly what happens, the quantum entities seem to know we are watching them and they adjust their behavior. They seem to know the set up of the experiment and how we plan to monitor it. If we have the detectors on, the light ends up in blobs on the screen, if we have the detectors off, it creates an interference pattern."[16]

"That's spooky," said Carolyn.

"That's quantum mechanics," said Uncle Jack.

"Generally when we talk about quantum mechanics we talk about how our world is not predetermined. There are groups of people who believe everything is ordained. They claim, "it's all in the cards." What that expression means is that our life and events are all linear. If you could see the future from your perspective today, there would be only one line or series of events that could possibly happen to you. What quantum mechanics says is that the world is not predetermined and it has randomness to it."

"If we get back to our Bohr model for a moment, quantum physics changes it. Instead of electrons orbiting the nucleus of an atom the electron is in an orbital. The electron is not at any certain point. It is constrained within a certain region, which is called the orbital. The electron can switch from one orbital or energy level to another instantaneously. It appears in one spot and then in the next instant is at another energy level. This is what is described as a quantum leap."[17]

"I guess that's where that expression came from," speculated Brian.

"Electrons are also affected by what is happening in events in other places. Non-locality is the term used to describe how events in one location affect local events instantaneously. Einstein called this 'spooky action at a distance'.[18] It ties into your comment earlier Carolyn. How is this possible if Einstein believed that nothing could travel faster than the speed of light?"

"Here is another interesting story. Cleve Backster was one of the key inventors of the polygraph test, which is commonly known as the lie detector. He is a master trainer of the application of this test. He has also used this device in very different ways, which illustrates how local events can be affected by non-local events."[19]

"Cleve had a test subject rinse his mouth with a weak saline solution. Electrodes were placed into the solution, which was then attached by electrodes to a polygraph machine. The actual subject was placed in another room and was not hooked up to the device. When he was provided a provocative magazine and he looked at a picture of a nude woman, the polygraph readings went off the chart. When the doctor involved in the experiment removed the magazine the polygraph results returned to more average readings. However, when the doctor provided one more look, again the polygraph readings went off the chart."[20]

"That's incredible. Are you saying that the saline solution contained cells from the subject's mouth and they were reacting to his actions in the other room?" asked Tom.

"That's correct Tom," answered Uncle Jack. He went on, "he did another experiment on the television show 'That's Incredible', wherein he took a single cell from a woman's mouth and connected it to an electroencephalograph (EEG) which produces similar results to the polygraph. She was then asked to walk through a rough section of town at night while a T.V. crew followed her. For the viewers at home, one half of the television screen covered her actions while the other followed her EEG readings. When a person accosted her, the EEG recorded her emotional reactions. Remember that the EEG machine was three miles away, hooked up to a single cell from her mouth."[21]

The group took a moment to let this new information sink in. The only sound that could be heard was the fire crackling.

Carolyn said, "This spooky action at a distance is really amazing. I can understand why Einstein called it spooky. It makes sense, but is weird when you think a few cells or even a single cell taken from you actually is still connected to you in some way."

The proof of these experiments created excitement within the group and they quickly exchanged whispered comments that this was precisely the stuff that drew them to these discussions night after night. Uncle Jack had them in the palm of his hand.

"In the physicists world, John Bell constructed mathematical proof that 'instantaneous–action-at-a-distance' could occur in the universe. In 1982 French physicist Alain Aspect conducted actual experiments that proved this. The net of this experiment is that two paired electrons appear to be separated, theoretically across the universe. No matter what the distance may be, it is as if they are tangled in some fashion that when one changes in direction the other changes instantaneously."[22]

"Now if these paired electrons were on either side of the Virgo Galaxy Super cluster, where light takes 100 million light years to travel across, this would challenge Einstein's theory about the speed of light. What does this mean?"

Brian ponders the question and replies, "Einstein's rule that nothing can go faster than the speed of light is not true."

"That is a logical answer Brian, but the truth here is that the pair of electrons are not separate objects, they are actually one whole, even though they are separated by tremendous space. So Einstein's theory is still valid."[23]

"We still don't have all the answers as to why our world is made up the way it is as described by quantum mechanics and it continues to amaze the scientific world. Despite the weirdness of quantum mechanics, it has illuminated rules, which the universe abides by. We just have to accept these rules, which are a mystery to us. By doing so, we have produced products utilizing quantum mechanics. We use these in our everyday lives, items such as lasers, CD's, DVD's etc.[24] For instance, you don't have to know how a DVD player works, it just does. You accept it as it is and use it. Many scientists delve into how to leverage quantum mechanics as they discover new applications and others focus on understanding more about its mysteries," said Uncle Jack.

They all wondered how this information, as interesting as it was, would apply to them.

"Quantum mechanics sure is weird Uncle Jack," said Carolyn.

"If you could have a nickel for the number of people who have responded in the same way, you would be a very rich girl Carolyn" replied Uncle Jack.

With that, Uncle Jack stood up and extinguished the fire with a bucket of river water and they all left for their beds as thoughts of quantum mechanics danced in their heads.

Key messages or ideas to take away from this Chapter:

1. Quantum Theory removes the deterministic vision of the world. There are always many possible outcomes.

2. Through quantum mechanics we learn that there are no basic building blocks that exist to create material objects, just energy and information.

3. The universe we live in is one big eco-system.

4. If you can remember one thing about the quantum world, remember the double slit experiment and how individual photons or electrons one by one build the interference pattern like a light bulb would do. And, if you try to observe the individual photons or electrons the interference pattern disappears. The point is that the mere act of observing affects everything.

5. While nothing can travel faster than the speed of light, as Einstein predicted, there is the instantaneous ability to have spooky action at a distance.

6. We use products such as DVD's, CD players, lasers that all are built on the principles of quantum mechanics.

7. Quantum mechanics takes us out of the Newtonian world into the world of magic.

4 $E=mc^2$ & Relativity

There are two ways to live your life. One is though nothing is a miracle. The other is as though everything is a miracle.

—Albert Einstein

The bonfire was the brightest light on the shore of the river. The sound of singing could be heard across the water as the activity on the river had slowed down for the day. As the singing tapered off, many people headed back to their cottages leaving Uncle Jack and the core group of his bonfire students.

"Tonight I want to delve into the finer points of energy. The world that Albert Einstein explored with his famous equation $E=mc^2$," said Uncle Jack.

"You have been referring to $E=mc^2$ so many times Uncle Jack that I have been waiting for you to tell us what it means," said Carolyn.

"Yes, this is the famous equation I hear about every now and then, but I never remember what it means. Doesn't it have something to do with the speed of light?" asked Tom.

"It does have something to do with the speed of light and it also has a lot to do with energy and mass. In fact, it is an equation that shows how they are related. So let me start this story by saying energy never disappears it just changes form, this is defined in the equation $E=mc^2$. In fact, energy equals mass multiplied by the speed of light squared." Uncle Jack continued, "Energy vibrates at different rates. Sometimes energy is in the form of a material object; sometimes it is what we consider pure energy, like heat or light. Energy can be locked into an object, as in a lump of coal, which is potential energy. When the lump of coal is lit, it burns and unleashes its energy into heat, which can help us transform other materials. If coal is burned in a blast furnace in a steel mill it helps us produce steel from raw materials such as iron. Also when coal is burned in thermal electric plants it produces electricity, which we use in industry and our homes. So mass can be transformed into energy. How about the other way? Can energy be converted to mass?"

All eyes were looking skyward, thinking about a possible response, but Uncle Jack responded quickly saving the group from having to supply an answer.

"One way it happens is through manifestation. Let's discuss that briefly. Manifestation starts with a thought, which is pure energy. If the thought were in the mind of an artist, it could be converted to a material object such as painting or a sculpture. In any event, I'm getting ahead of myself again, so let's return to the present moment and $E=mc^2$."

"The equation $E=mc^2$ is perhaps the most famous equation known to man. Albert Einstein developed it at the turn of the 20th century. What Einstein postulated in his theory of relativity is that mass will convert to energy at the speed of light squared. C is defined as the speed of light, which equals 670 million mph or commonly referred as 186,000 miles per second. So therefore C^2 is 448,900,000,000,000,000 mph.[1] As any object approaches the speed of light certain things happen. These events may seem rather strange to us, however, as we've discussed throughout our bonfire chats our perception of life and the universe we live in, is only that; *a perception*. In reality, it is much more magical. If we embrace this thought, we will be able to make quantum leaps in our lives. Again, I will talk about that during our bonfire chat on manifestation."

The first few times the group listened to Uncle Jack, they found he used words and terms that sometimes went over their heads. The more time they spent listening, learning and asking questions, the more they began to understand those terms. So his use of the words; perception, quantum leap and manifestation with respect to how they live, started to become part of the vocabulary of the group.

"So how does E=mc^2 change our perception of the world?" asked Carolyn.

"Let me see if I can answer that Carolyn. Let's talk about how energy and mass relate to C^2, the speed of light squared. Einstein's special theory of relativity causes us to challenge our perceptions of reality. Imagine for a moment that a spacecraft takes off from earth, accelerates using booster rockets and approaches the speed of light, Einstein postulates that two key things will happen:

1) Time will slow down relative to time outside the spacecraft i.e. back on earth;

2) The spacecraft and its passengers will increase in size infinitely as it approaches the speed of light."[2]

"How is that?" inquired Brian. "I think you need to explain those two points."

"As you get closer and closer to the speed of light, time on the spacecraft would get slower and slower. For example, one year on earth could be minutes on the spacecraft. And secondly, the spacecraft and all its passengers would keep expanding infinitely. They wouldn't feel any changes in time or in their size. The change, for example, would be relative to that of the time and size of the inhabitants of earth."[3]

"That sounds a bit like a science fiction movie," chuckled Brian, who was not fully grasping this concept.

In an effort to explain, Uncle Jack continued, "Einstein believed that you could never actually reach the speed of light because matter would change into energy. That is what the equation $E=mc^2$ means. Mass and energy are equivalent. C^2, which is the speed of light squared, happens to be the conversion factor. This is like the factor 2.54, which converts inches to centimeters. If a measurement is 10 inches long, we multiply it by 2.54, which equals 20.54 centimeters. So 10 inches is equivalent to 25.40 centimeters. When mass or energy reaches C^2 they can be converted into each other. So, if the spacecraft actually hit the speed of light the spacecraft and its inhabitants would explode into energy."[4]

Tom responded firmly, "Okay I've got it. Sounds like shades of quantum mechanics, but I get the point of the equation now."

The others nodded in agreement.

"The Special Theory of Relativity was proposed by Einstein in 1905. This theory predicts that the laws of science are the same for all observers no matter how they are moving, excluding the effects of gravity. It provided birth to the equation $E=mc^2$. Then in 1915, Einstein finally explained how gravity impacts his theory of relativity through the General Theory of Relativity. Gravitational effects were described in terms of the curvature of four-dimensional space-time."[5]

"Let's go back to our example of the spacecraft. You would think that as you approached the speed of light you would be able to throw on more energy, however, by adding more energy it gets converted into more mass. It does not get converted to more speed, as you would think, because you can never go faster than the speed of light. And why is that?"

"Because the spacecraft would expand to infinity as it approached the speed of light and as it hit the speed of light it would explode into energy," answered Tom.

"Bang on. It would become pure energy," said Uncle Jack.

"Again, if you could view the spacecraft and its passengers from the outside it would be growing larger. However, the passengers would not realize any change in themselves, as everything in their environment would remain constant."

"How do we know this Uncle Jack?" asked Brian. "I know that the fastest airplanes don't go very fast relative to the speed of light."

"I like your use of the word 'relative' Brian, it is appropriate in the theme of this discussion. Would it help you to know that we have scientific proof of this? The phenomenon that Einstein described actually occurs in particle accelerator labs outside of Chicago and at CERN in Switzerland. As the protons are flung around in the particle accelerator, the protons swell as more power is pumped in. At the speed of 99.9997 percent of the speed of Light (C), the protons end up 430 times larger than their original size. So much power is drained from the nearby electricity stations that the main experiments are often scheduled to run late at night, so that nearby residents won't complain about their lights dimming."[6]

"Okay, I'll buy that proof," said Brian.

"I knew you'd be tough sell, so I pulled that one out just for you Brian. Let's deal with the effects of time related to the speed of light. Einstein called this time dilation. If you could travel at near light speed, time would slow down drastically relative to time on earth. It is kind of like *time travel*.[7]

"Einstein had a great example of this he called the *twin paradox*. Imagine twin sisters aged 10. One sister takes a trip into space at 99.5% of the speed of light for 5 years. Upon her return to earth she would find her earthbound sister had aged by 50 years while she had only aged by 5 years. In other words they now have a 45 year age gap."[8]

"Wow, now we are talking sci-fi and it sounds like that old movie *Back to the Future*. You sure would be able to tell the difference between those twins," quipped Brian.

"How can we prove this one Uncle Jack? It's not that I completely doubt what you are saying but you've pretty much given us a proof statement for everything else and I'm interested to see if you have one for this," inquired Carolyn.

"Great question Carolyn. During a Space Shuttle mission

in 1985, a highly precise atomic clock was utilized to run an experiment called NAVEX. Despite the fact that the space shuttle orbits at 1/42857th of the speed of light, they measured a slowdown of 0.000,000,000,295 seconds for each second of flight. This was almost exactly what Einstein's formula predicted," Uncle Jack responded.[9]

"I would imagine that the fact that the airplane is traveling at a fraction of the speed of light is the reason there is only a few billionths of a second difference," stated Tom.

"That's correct Tom," said Uncle Jack.

"I love this stuff," exclaimed Tom.

"Okay, let's examine some of the less spectacular aspects of relativity, but nonetheless an important facet of Einstein's perspective of gravity. The other night we talked about gravity and Newton's perspective, now let's look at how Einstein applied this to his theory of relativity and what it means to us. Einstein explained that the earth doesn't move in a curved orbit around the sun, it actually follows the nearest thing to a straight path in a curved space called a geodesic. The easiest way to describe this would be to visualize how pilots navigate airplanes along 'great circles' around the earth when they travel between two points. The geodesic is the shortest distance between two points. So the airplane traveling between two points or the earth revolving around the sun follow straight lines in four-dimensional space, however, we see them traveling in curved paths in our three-dimensional space perspective. Both Newton's and Einstein's theories are almost identical when they predict the orbits of all of the planets, with the exception of Mercury. Every 10,000 years Mercury's long axis of the ellipse will rotate by one degree and this effect was one of the first confirmations of Einstein's theory. In fact, other deviations in the orbits of the other planets in our solar system have been established with radar and have confirmed Einstein's equation."[10]

"Light also follows a geodesic line rather than straight lines as predicted earlier. When light comes from a distant star its actual location is not where it appears. The gravitational forces of the sun affect the light from distant stars, or more succinctly, the space-time curvature created by the mass of the sun affects the light. The sun ends up deflecting the light on a different angle, so that the star isn't actually where we think it is."[11]

"Let me paint you a picture of this."

"Please do Uncle Jack," chuckled Carolyn.

"Imagine four people holding each corner of the blanket you are lying on. Now imagine someone putting a bowling ball in the middle of the blanket. The bowling ball on the blanket is the equivalent of how the sun distorts space-time. Now, if you rolled a baseball across the blanket with enough force that it could get to the other side it would not go in a straight line. The weight of the bowling ball in the middle of the blanket would force the ball to deflect on an angle because the blanket is not perfectly flat. Light from a distant star would react the same way given the space-time distortion created by the sun. It's as if the sun's mass affects the fabric of the universe and distorts space-time. This effect was proven in 1919 by astronomers who carried out experiments by taking photographs of the heavens during a solar eclipse. As the earth orbits the sun, different stars are seen as being closer to the sun at different times of the year. However, if the light from distant stars is near the sun, we can't see these stars because the sun overpowers their light. In the moment that the sun's light was blocked by the moon in the 1919 eclipse, a star appeared in a different position than expected, thus proving Einstein's theory about gravity."[12]

"Furthermore, as described by the General Theory of Relativity, time runs slower near massive heavenly bodies, for example the earth, sun and black holes. As light travels upward in earth's gravitational field its frequency reduces. Imagine for a moment, waves blowing across the surface of the river on a windy day. The mass of the earth increases light's frequency and the crests of the waves are closer together. The further you get away from the gravitational effects, the frequency decreases and the crests of the waves are further apart. To someone observing earth from a great distance it would appear that everything was taking longer to happen. Which means that time would go faster at this great altitude relative to time on earth. In 1962 this prediction was tested with the aid of two atomic clocks, one on the top and one on the bottom of a water tower. The clock at the bottom was found to run slower."

"In actual fact the time distorting effects of relativity come into play with the use of the U.S. Department of Defense's GPS navigational satellites. The signals that they beam down are constantly shifted out of sync by the time distorting effects of relativity. These satellites travel so fast and at such a great altitude that time is different relative to time on earth. As they soar high above the earth, these satellites have to be continually programmed to correct the effects of relativity.[14] If one made a miscalculation, the readings from these satellites could be off by many miles,"[15] concluded Uncle Jack.

"The slowing down of time and the conversion of energy into infinite mass as you approach the speed of light all seems hard to believe, however, we know through our examples that it is true. How could this be possible and what does this mean? Again, it tells us that the universe is a magical place. What we see and touch isn't necessarily all there is. The Newtonian worldview is dead."

"Uncle Jack, the interesting thing is that the Theory of Relativity and quantum mechanics are all very unusual when compared to our Newtonian world as you say, but they all have practical applications. And of course in our day-to-day lives, we would never know that. Thanks for telling us about quantum mechanics, $E=mc^2$ and relativity," said Tom.

"It is my pleasure Tom. I believe you need to know this information. It's important to me that you know that the world isn't just what you perceive and that the universe is made up of rules that dispel the Newtonian worldview because I want you to know that there is magic out there and the world and our universe isn't just what you see, hear, taste and touch. You can use this magic to create your lives. I know I'm repeating myself but I am so passionate about this I just can't help it."

"There is no disputing that fact," exclaimed Brian. And at that moment all that could be heard was a roar of laughter coming from the group on the beach. Even Uncle Jack was chuckling.

As the group's laughter subsided, he continued. "Currently we do not have the technology to travel at speeds that approach the speed of light, and as Einstein described we cannot go faster than the speed of light. This begs the question, will we ever be able to travel great distances in space, to other star systems, galaxies etc? Time will tell...pun intended. In 1935 Albert Einstein and Nathan Rosen suggested that we could take advantages of a warp in space-time and that within space-time there exist wormholes that are thin tubes of space-time, which connect different regions of space.[16] Perhaps you could travel to Alpha Centauri through a wormhole and get there much faster than 4.7 light years, if it were possible to travel that fast to begin with."

"This is one theory that I do not have any proof for. We will not be able to conclusively test all of the theories that Einstein predicted, perhaps at least in our life time. We just don't have the technology yet."

"Sadly, we have been able to test the energy locked in material objects through atomic explosions. Atomic energy is a wonderful thing if it is used for creating electric power, unfortunately it has also been used for destruction. The development of the atomic bomb conclusively proved Einstein's General theory of relativity. In layman's terms, what happens in a nuclear explosion is that material collapses upon itself and electrons and neutrons are exchanged at the speed of light and this initiates a chain reaction. This chain reaction unleashes tremendous energy, which results in a nuclear explosion. Today we can easily unlock the energy stored in uranium to generate nuclear energy, however, nuclear energy is locked into every material object. If you could extract the energy stored in a single page of a book, which weighs only a few grams, it could erupt with more energy than what is produced by a large power station. A single pound of mass in the equation $E=mc^2$ (1 lbs = 1 lbs x 448,900,000,000,000,000 mph) would produce more than 10 billion kilowatt hours of energy which is equivalent to a huge power station by today's standards.[17] While this proves to be true as described in Einstein's equations, we don't have the advanced methodologies to put this in practice today."

"Through the use of uranium scientists were able to prove Einstein's $E=mc^2$ by unlocking the atom. Uranium was the first material used because it was, by nature, less stable. Scientists were able to push extra neutrons into the nucleus making the nucleus wobble and then explode. The addition of the extra neutron made the glue of the strong nuclear force give way and the inner parts flew apart with wild energy. If you did this on a very small scale, you could check the weight before and after the addition of the additional neutron and you would find that the some of weight seemingly disappeared. The mass that disappeared was converted into the energy. It actually escaped at the high velocity of C^2.[18] The equation $E=mc^2$ coupled with Einstein's Special and General Theories of Relativity changed the way we see things."

"It is unfortunate that such a great discovery could be used for destruction," said Carolyn.

"There are infinitely more positive applications of this Carolyn and I agree there is a very dark side that mankind must figure out how to manage."

"Let's reflect for a moment on the thoughts of people who have lived before us. You don't have to go back very far, consider the days before Columbus discovered America. People thought the world was flat and that the sun revolved around the earth. Galileo was imprisoned for thinking that the earth revolved around the sun. Listening to radios, music on walkmans, watching television, talking on telephones or mobile phones were improbabilities. Just imagine what might be theorized and developed over the next thousand years. We must be open to the possibilities. Life isn't just what we see and touch. I would like you to consider that if a space craft could travel at 99.999% of the speed of light and we know that the occupants would expand infinitely and that time would slow down infinitely relative to our size and our sense of time on the earth, such as in the 'twin paradox' example, then we really do have a magical universe," concluded Uncle Jack.

With that statement the questions ceased. Tom and Brian helped Uncle Jack put out the fire. The group collected their things and headed for bed.

Key messages or ideas to take away from this Chapter:

1. $E=mc^2$ – Energy equals mass at the speed of light squared.

2. Energy is locked in every material object. In fact energy and mass are the same thing, they appear to us in different states like water and steam.

3. At the speed of light every material object converts to energy.

4. To date, Einstein's theory holds that nothing travels faster than the speed of light, which is 670 million miles per hour (186,000 miles per second).

5. Einstein proposed the Special Theory of Relativity in 1905. This theory predicts that the laws of science are the same for all observers no matter how they are moving not including the effects of gravity. It provided the birth to the equation $E=mc^2$.

6. Then in 1915, Einstein finally explained how gravity impacts his theory of relativity through the General Theory of Relativity. Gravitational effects were described in terms of the curvature of four-dimensional space-time.

7. Here is an example of the Special Theory of Relativity. If a spacecraft takes off from Earth and accelerates using booster rockets and approaches the speed of light, Einstein postulates that two key things will happen:
    § Time will slow down relative to time outside the spacecraft i.e. back on earth (proof – atomic clock experiments in the space shuttle and the current use of GPS satellites)
    § The spacecraft and its inhabitants will increase in size infinitely as it approaches the speed of light.

(Proof – proton expansion at near light speed in particle accelerator experiments.)

8. $E=mc^2$ and Relativity evoke thoughts of magic. The best example is the twin paradox thought experiment.

5 *Are Solid Objects Really Solid?*

The noblest pleasure is the joy of understanding.
—Leonardo da Vinci

Another hot and hazy day had come to an end. Despite the fact that the sun had set, the air was still and heat continued to radiate from the stones on the beach. The bonfire was lit and the group was lying back gazing at the universe above.

Uncle Jack started the evening's discussion by asking the group, "Are material objects solid?"

Tom responded quickly, "If I remember the quantum mechanics chat, I believe they are made up of atoms which in turn are made up of non-material things that we call energy and information."

"Great answer. Anything to add to that?" asked Uncle Jack.

"Yes," replied Brian. The atoms are mostly void of anything, there is a tiny nucleus in the middle and then swirling around are electrons and other things like bozons and quarks" said Brian.

"I love it when I'm surrounded with such great intelligence," exclaimed Uncle Jack with a smile as he gave Brian a pat on the back.

"Having said that, is the cliff behind us solid? Or is the bedrock under the thick layer of sand and shale solid?" There was a pause and then Uncle Jack said, "Okay, I'll answer that, but I think you already know that answer. As we have discussed it looks solid, however, through our science lessons we know that it isn't. Light and sound are unable to pierce through thousands of feet of what we think is solid bedrock. The operative words are 'what we think' is solid. Yet, objects called neutrinos are able to do this. Neutrinos are continually bombarding our planet. Billions of neutrinos are passing through your thumb while I finish this thought.[1] It certainly begs the question, what are these things?"

"Is this something I should be concerned with Uncle Jack?" asked Carolyn.

"No Carolyn, neutrinos have been bombarding the earth since its creation. Neutrinos are either waves or actual particles. They have characteristics of both, which sounds a lot like how we describe light. Scientists cannot confirm which model describes them accurately. As you know, if they have mass they are considered a particle. Otherwise, they are considered a wave, which again can be compared to light, x-rays and UV light etc. The current thinking is that they may have mass, but it is almost zero."[2] Uncle Jack paused for a moment, letting that thought sink in.

"What we do know is that neutrinos originate from the stars. The neutrinos that bombard our planet come from our sun. Enrico Fermi theorized that neutrinos existed. It was twenty-six years later that Wolfgang Pauli finally detected the existence of neutrinos.[3] The sun is a nuclear generator that runs on nuclear fusion. Neutrinos are the by-product of nuclear fusion. In the 1930's neutrinos were conceived as a possibility, a hypothesis. Scientists were studying radioactive beta-decay. Beta decay occurs when a neutron converts to a proton and when that occurs an emission of an electron occurs. When scientists measured the energy that was produced during this conversion all the pieces did not add up. Something was missing. That is where Pauli explained that there was an invisible particle that was emitted with the electron carrying away the missing energy and momentum."[4]

"It is amazing that these scientists and physicists keep studying these subjects and finding these interesting things. I wouldn't know where to start," said Brian.

"As I recall Brian, before we started our bonfire talks, you didn't know much about the universe or anything about quantum mechanics or relativity. You listened and you asked a lot of questions, it would seem that the more you learned, the more you wanted to learn. This is what inspires the great scientists and physicists of our world. They never stop asking questions and searching for answers."

Uncle Jack continued, "Since this discovery there have been many experiments conducted wherein scientists collected neutrino hits in tests above the ground. However, since other particles or waves were similar to neutrinos it was difficult to discern whether the scientists were actually collecting neutrino hits."[5]

"In order to consistently study neutrinos scientists created neutrino observatories. About 200 miles from here in Sudbury, there exists one such observatory that is located 6,800 feet below ground, under solid bedrock in a mineshaft."[6]

"That's a weird place to have an observatory," interjected Tom.

"Well, not if you are trying to observe neutrinos. Neutrinos move at the speed of light and you may remember that the speed of light is 670 million miles per hour. They can move at this rate of speed through solid matter for up to a 1000 years before they would come to a stop."[7]

"Wow," said Carolyn.

"The observatory in Sudbury was specially designed to detect the presence of a neutrino. In a large space 6,800 feet below ground there exists a ten story sphere filled with heavy water. The sphere is lined with 13,000 light detectors that can measure as little as one photon of light. A photon of light is light energy that is emitted by an atom during a transition from one energy state to another. An average 100-watt light bulb that you would use in a home emits 12.5×10^{20} photons per second. As you might imagine, in order to detect one photon, these light detectors must be very sensitive."[8]

"A neutrino is created in the sun and bombards the earth at light speed passing through our atmosphere and through the bedrock and into the sphere in the mineshaft. When it hits the heavy water in the sphere in just the right way (how it strikes the water molecules), it produces a burst of blue light (a photon of light), which is detected by the sensitive light detectors. Heavy water is used because it has an extra hydrogen atom and therefore there is more of a target for the neutrino to hit."[9]

"Scientists determined that all of the neutrinos that are directed at earth do not all reach the surface. Where were the missing neutrinos? By measuring the neutrinos in the Sudbury observatory they were able to determine that there are different types. It was determined that there are 6 different types of neutrinos and that the sun only emits one type. So at some point the sun's neutrinos are converted into 5 other types before they hit the surface of the earth."[10]

"All of this is fascinating, however, how does this relate to the original question with respect to solid objects? It does prove to us that what we think is solid, is not. It is an illusion. Trillions and trillions of neutrinos are passing though our bodies and directly through the planet continuously. We have grown accustomed to thinking that solid bedrock is in fact solid and we know that mobile phones and radios do not work effectively through some of the massive concrete buildings let alone through mountains or bedrock. The fact that neutrinos can move through these 'solid' objects forces us to change our perspective."

"So you might ask, how exactly do these neutrinos slip through? They are able to do this because they are almost at zero mass.[11] They are able to go between spaces or voids as we discussed before. This helps prove that matter is not solid at all, it is only an illusion."

"Let me get this straight Uncle Jack, what you're saying is that not only do I have illusions in my mind...I have neutrinos passing through there too," chuckled Brian.

"If I had never told you about neutrinos, you would have thought it impossible to conceive of something going through perceived solid objects. If you did think something could pass through things you might expect to find holes. The billions of neutrinos that have passed through your thumb in the past few seconds have made no apparent damage to your body or anything else around you."

"I'm glad for that," said Carolyn.

"The study of neutrinos helps scientists understand the sun, stars and our universe. However, the purpose of this story about neutrinos is important to convince you that what you think is a material object is merely an illusion. The purpose is not to make you disbelieve that objects are real. What is important is that underlying any material object is energy and information and no real material object at all. And therefore everything really is energy and information and this includes sound, material objects and most importantly thought. It is all one and the same thing. This all leads me to my discussion about manifestation, however, that is for tomorrow night. We have reached a point where we can talk about manifestation as you have now graduated from pre-school bonfire chat class into kindergarten bonfire chat class."

And with that announcement Uncle Jack took the pail of river water he usually uses to put out the fire and surprised the group by pouring it on them, "congratulations." he shouted.

As the night was still so hot and steamy he jumped in the water, clothes and all. The rest of the group, stunned and already soaked, jumped up and chased him in. They laughed and pretended to get even with Uncle Jack by splashing him as the fire and full moon lit their way. It seemed like a fitting end to their pre-school graduation ceremony.

Key messages or ideas to take away from this Chapter:

1. Neutrinos travel at the speed of light. They can travel through solid matter such as lead at the speed of light, eventually stopping in 1000 years.

2. Billions of neutrinos are traveling through your body every second.

3. All material objects are made of energy and information as learned through quantum mechanics. There are no material building blocks at the quantum level.

Part II

Leverage the Magic

6 *Manifestation*

All that we are is the result of what we thought.
—Buddha

Change your thoughts and you change your world.
—Norman Vincent Peale

There are times when the beauty of Pine Ridge is lost in the daily hubbub of activities and there are other moments when one is in awe of its spectacular setting. Looking from high on the cliffs, the vision across the river cannot help but be forever etched in the mind of the observer. The glorious vista takes on the breathtaking beauty of the setting sun, which leaves streaks of brilliant orange. Down river, the little white church dots the countryside like a perfect sugar cube, nestled in the hills framed by the majestic forest clearly outlining a picture of the beauty and perfection of nature.

This is a time when the river begins to quiet down and its waves begin to smooth out to create a giant mirror in which to reflect the sunset. At the river's edge, the activities also settle down as the residents are in the process of doing what is meaningful for them at this special time of the day. Some are shedding their t-shirts and shoes for that last dip in the alluring waters. Others are walking along the beach, enjoying the company of neighbors or are perhaps meandering on their own in a meditative manner. Adults are supervising the youngest children as they explore the boundaries that their parents and the outdoors provide. And, lastly, there are the fishers, who cast their bait from the shoreline and the docks contemplating the possibilities with every turn of their reels.

As Uncle Jack is preparing the evening's bonfire, Tom helps him arrange the logs and the kindling to ensure that all will enjoy the warmth of the flames when the sun finally takes its evening rest.

Throughout his childhood and early adulthood, Tom experienced many sunsets at Pine Ridge and very much enjoyed the company of Uncle Jack. His return visit to Pine Ridge was more than a vacation from work. He had reached a point in his life where he was unsure of his next steps. He had always dreamed of the success he would achieve in his life, however, those dreams were not being realized. He had dreamed that he would be married with children, living in a beautiful home and perhaps owning a cottage in Pine Ridge. His objective during this time away from work was to return to his parent's cottage to contemplate his future and to seek the guidance of Uncle Jack.

As Tom gathered the kindling and expertly positioned it among the logs that Uncle Jack had arranged, he began to tell Uncle Jack about his dilemma. He spoke quietly so that no one else could hear. "Things are not working out as I planned. I seem to struggle to pay my bills, let alone reach some of the lofty goals I've set for myself. I don't know if I enjoy my job or if I'd ever enjoy any type of work. I just don't know how I am ever going to be a success the way I define it."

Uncle Jack listened quietly to the tale that Tom spun and every now and then he replied with an, "I see" or with "yes, I understand." When Tom came to a long pause, he told him the following. "I'm thrilled that you are here, we have all missed you over the years. I would like to talk to you about your confusion with the life you want to live and the life you are living. Manifesting is our topic this evening. I think you will find it helpful."

"What exactly do you mean by manifesting?" asked Tom.

"Tom, manifesting is about taking a thought and then making that thought appear in the physical world."

"This is what you have been leading up to in our previous fireside chats, isn't it?"

"Yes," replied Uncle Jack.

"So I guess what you mean by manifesting in this instance is goal setting. Like the type of job I wish to have or the kind of income I would like to earn. Or perhaps thinking even larger that this, to the kind of life I'd like to live?"

"You are right on." Uncle Jack answered, "Before anything occurs a thought must have preceded it, therefore, there are no accidents, including the thought you had that lead you to return to Pine Ridge for your vacation. Anything that has ever been invented by humans began as a thought, a mere possibility. This shows us how powerful thoughts can be."

"Isn't it a coincidence that I received your invitation to the bonfire lessons when I needed it the most," said Tom.

"Yes, it sure seems like a lucky coincidence Tom. Perhaps it's best described as a synchronicity," replied Uncle Jack.

Tom quietly contemplated that thought as he sat down by the fire and watched the sun's final rays reach the top of the cliff behind him.

Dusk was now settling in. Fireflies danced in the air as the firewood began to crack and pop, sending sparks in the air. Faces were becoming muted as the light slowly faded. The campfire songs started up with a few lone voices, but before long the group was swaying to the rhythm as their collective voices carried down the river on the back of the evening breeze.

As the evening wore on, the group began to dwindle until only the core group remained. They lay on their backs, hands clasped behind their heads gazing upwards, occasionally noting a shooting star or a passing satellite.

As the evening's breeze pushed small waves hypnotically against the beach they found themselves slipping into their own thoughts, moving into a somewhat trance-like state.

They knew the balance of the week was to focus on leveraging what they had learned in the past few days. As they reviewed the lessons, the irrefutable facts of science and nature were astounding.

Carolyn thought about how magical the universe is. She had never really understood Einstein's work, let alone the complexities of quantum mechanics. Each evening as Uncle Jack progressively peeled back the layers of knowledge it created some interesting feelings. For Carolyn, it was as if the foundation of her youth had somewhat cracked. She felt some fear generated by her new level of understanding as she came to grips with the evolution of the universe and the magic described through relativity and quantum mechanics. The safety of her environment in Pine Ridge, Uncle Jack, the ability to ask questions and to see the same curiosity in the other two students gave her the ability to overcome the discomfort of her new worldview. While the magic provided freedom she found it interesting that it also generated some fear within her. It was a new perspective.

Brian on the other hand, had progressed with his thinking. His mind was expanding into thoughts of the future and he eventually broke the silence.

"One day I want to create the world's best snowboard park," he announced as he lay there dreaming about the future, seeing the silhouette of the Gateneau hills in the distance. "It would be designed for people of all abilities. There would be rugged terrain parks with colossal jumps and rails for the experienced snowboarders, with plenty of space to handle hundreds of people. There would be coaches available to help those who wanted to improve their boarding skills. I would set up another area with coaches to help the beginners learn how to snowboard and do the basic tricks so that they wouldn't interfere with advanced boarders. That way everyone could progress at their own speed."

"How could you be thinking of winter at a time like this?"

inquired Carolyn. They all laughed.

"Actually that is a great vision Brian, I know it is something you would really enjoy doing," said Uncle Jack.

"It would be great. I love designing things and I love snowboarding," said Brian. "The problem is that I haven't got an idea about how I would ever accomplish something like that."

"Brian, you have taken the first step," answered Uncle Jack.

"What's that?" asked Brian.

Uncle Jack answered, "The first step in making anything happen is a *thought*."

At that point Tom jumped in and explained to the group all he knew about manifesting from his prior discussion with Uncle Jack.

Uncle Jack continued, "Manifesting your destiny is something only *you* have control over. And, believe me, you do control it, no one else does. Things do not happen by accident." He repeated the lesson he had given Tom. "You don't just get to the age of 70 and say you made it there by accident. You have to plan it. Some people make very specific plans while others just have thoughts in their heads and don't even realize they are designing their own roadmap for the future. Let's get back to the basics so that we really understand what manifesting is about."

"Think of anything you have in your cottage, home, school, or at work. People manufactured these goods, such as: computers, televisions, radios, pencils, papers, desks, chairs, snowboards, basketballs etc. Each one of these items was a thought in someone's head before it was designed and materialized in the physical world. What I'm introducing here is the concept of personal magic. Magic, that you can create. Where a thought materializes into a physical object or action."

"What we talked about in our earlier discussions was amazing stuff: Einstein and his theories of relativity, $E=mc^2$, quantum mechanics, the universe we live in. Through these discussions we opened our eyes to the magical environment we live in. Manifestation takes it to another level. It's about how you can transform your life. Manifestation is truly the magic that *you* create, in your own special way. This is a special gift that we all have and you must trust that you have the power to manifest things in your life."

"I want to manifest something exciting." said Carolyn.

"The funny thing is that most people don't realize that magic is occurring in their lives every day and that they are living the life of their dreams. If your life is not what you want it to be, change it. Dream new dreams. Imagine your life as you want it and then make it happen. This may sound too easy but that is all you need to do. Many people are fearful of ever attempting their biggest dreams because they are afraid of failure. Herein lies the key. If you think of failure, guess what you are going to manifest? You got it, *failure*. You will manifest what you are focused on. My point is, be selective in your goals for the future and be careful what you focus on. The golden rule is: focus on what you *want*, not on what you *don't want*. If you find yourself thinking of thoughts of failure, say to yourself, *cancel that thought* and think about how you are moving towards your intended goal. Sometimes you are only slightly improving your position, but nonetheless, you are moving towards it. This is very important to manifesting."

"Along the path you may find that what you really thought you wanted is not the case and then you can change your goals and intentions. Don't confuse this with failure because the universe provides many alternatives. As one door closes another one opens. So, tonight I want to discuss the possibilities, tomorrow we will focus on the obstacles and challenges that we always face."

Uncle Jack then changed the pace for a minute and brought the conversation down to a different level and started to explain to the group how they experience manifesting on a daily basis. "You know, all of you manifest everyday. Today you manifested the plans that you had set the day before or earlier this morning or perhaps even before you came up to Pine Ridge to spend the summer."

"But we didn't make anything today," commented Carolyn.

"Carolyn, when you manifest something it doesn't have to be a material object. Manifesting is making anything happen that begins as a thought. For instance, I overheard a group of you planning your canoe trip to the falls yesterday. When I saw your smiling faces and listened to you sharing your stories on the beach this afternoon, it sounded like the plan came together quite nicely."

"Yes, we had a blast today," said Tom. Although during our 'up-river' morning canoe the complain-o-meter was hitting ten on a number of occasions. I'd say about three of the eight canoes were lagging behind. Most of their paddling energy was spent flapping their jaws instead of their arms. I kept on hearing excuses that it was too hot and it was too hard to paddle with those old paddles…the whining went on and on."

The group erupted into laughter as they thought about their experiences of the day.

Uncle Jack roped the group back into the discussion as the laughter died off, "manifesting your destiny is just a longer version of manifesting your day. We all think and discuss visions of the activities that we want to pursue. Let's look at an example of manifestation and how we are integrated with others like the eco-systems we mentioned the other night."

"Imagine you have a pencil and a pad of paper. In your mind you have a thought to either draw a picture or write a letter. You start with the raw materials; the only additional component you require is the thought of what you would wish to do. In fact, typically you have the thought first and then seek out the pencil and paper. If you didn't have the supplies you would go out to buy them or borrow them from someone."

"You begin with the intention to write a letter, with a pencil and a pad of paper you begin to express your thoughts. As you write, the thoughts come streaming out and creatively appear on the pad of paper. When you have completed your thoughts you have created a new material object – a letter. You have literally created something from nothing other than thoughts. This object never existed in material form anywhere in the world."

"Let me explain this further. Not only did you create the letter from thoughts, which are energy and information, you used the energy and information that was stored in the pencil and paper, which were manufactured by other people. Call that 'potential energy and information' waiting for the right people to use them at the right time. Going back further in the chain, other people invented pencils and paper for this reason; to write letters, or to put it more broadly, to communicate with others. Then the companies who purchased your specific pencil and paper thought there would be a market to sell them. So, you combined your intentions with the intentions of the inventors and the manufacturers of the products. All of this started from thoughts, which are energy and information. This is the simplicity of manifesting."

"You can take this example and use it to describe the basics of how anything is invented, manufactured or accomplished. For instance, take the example of putting man on the moon before the end of the 1960's, in this case they also leveraged the intentions of other people in a collaborative fashion to produce the Apollo mission. The interesting thing about the space program is how it subsequently provides new products and services for everyday uses such as the special plastics that were used in the space program or powdered fruit drinks etc. All of this is developed as part of the vast ecosystem that we manifest in. All of the intentions that we have in life propel our knowledge forward, individually and collectively for the good of all mankind."

"What about all the wars and terrorist activities that occur. They don't seem to be in the best intentions of everyone?"

asked Tom.

"Sadly there are a lot of negative intentions and events that occur in the world. It is part of the human struggle, wherein a company in the west attempts to put another company and its people out of business because they are a competitive threat. Or in a war torn divided nation in the Middle East or Asia where an individual drives a car bomb into a populated building, killing innocent people and destroying property. These are tragedies."

"It is very difficult to understand with our limited vision, but from every negative comes a positive. Perhaps stronger business rules are put in place to prevent companies from flexing monopolistic power, thus creating a business environment where small companies can grow and flourish, which in turn stimulates the economy by generating jobs and providing more wealth and optimism for its citizens. And in the war torn nations the tragedy of the innocent victims who were killed and the motivation of the violent protesters eventually are healed over time when an understanding is reached. All of the world watches and learns through these events. Time heals, people learn and good does eventually rise, however, it can take many years to reach this point."

"How about failure Uncle Jack?" asked Tom.

"Failure to reach a goal is a great learning tool. Remember, failure is only a step in the direction of your goal or intention. This is when most people quit. Have any of you ever made a New Year's resolution to start a fitness program, perhaps to lose weight and re-energize yourself?"

Heads nodded.

"Well, I'm not pointing a finger at anyone in particular because this happens all the time with intentions. People are obsessed with their enthusiasm for fitness for two weeks, a month, maybe more but slowly they find themselves exercising less often and then eventually stop altogether. At this point they have decided to change their plans. There are many different reasons, but in the long run other things got in the way, or should I say, became more interesting. Their desire for fitness wasn't enough to offset additional requirements at work, school or other social obligations. Not to mention the fact that it takes discipline and hard work to follow a fitness regime. So one might ask, is it bad to get sidetracked or to divert your attentions elsewhere?" questioned Uncle Jack.

"I am going to address challenges and obstacles tomorrow night, but I will reflect on some of these issues as we talk about manifesting this evening. For instance, plans in life change, one day it is important to pursue your specific dreams and goals and the next day your priorities change. The most important thing is that you set your intentions and then do something about them. In the course of pursuing your long-term dreams or your short-term daily activities, other opportunities present themselves as you focus and execute your plans. As you pursue your intentions, you open yourself up to a world of other possibilities that you may never have encountered if you hadn't started."

"As I mentioned earlier many goals are not realized because your interests change or because you fail at some point along the way. Failure is a perception. Failure is actually an opportunity for personal growth. Thomas Edison realized this as he went through the process to invent the electric light bulb. He tried 3000 different ways to make an electric light bulb. His last attempt worked. The other 2999 attempts were possible methods to make an electric light bulb, which proved false. He persevered and applied attention and hung on to his vision. We will talk more about this subject tomorrow."

"I have spent a lot of time over the years, changing my path" said Tom. "In hindsight I always thought it was a waste of time changing from one sport or hobby to another or from one job to another once I lost interest. I thought I didn't have any staying power. But now that I listen to you, it makes sense. Our desires do change. I appreciate your perspective Uncle Jack. I often considered that I failed to reach excellence. But you are right, I just got *turned on* to better and greater opportunities that interested me more. I suppose you might say I was 'growing' during the many transitions in my life."

The group paused for a moment as the fire crackled, and they reflected upon their own experiences.

"Let's take a look at the basic ingredients to manifesting. It's very simple. Are you ready?" asked Uncle Jack.

"Go for it," said Tom.

"Okay, I'll list them for you:

1. first you need to have a *thought*;

2. then you need to *intend* its outcome; and

3. you need to apply *attention* to it in order to manifest it.

You may have the greatest *thought* but you have no *intention* of making it happen. And this is okay because you are not going to act upon every thought. Or you may have a great *thought* with all the best *intentions* to make it happen, but then you fall short of your goal because you didn't apply any *attention* to it."

"The same sacrifices and focus have been applied by business people like Edison and Bill Gates. Also by spiritual people like Mother Theresa and Gandhi. They have all forgone things in their lives at some point to reach their goals. When they achieved their goals they also became abundantly wealthy in many ways, which allowed them to do other things in their lives that they never dreamed possible. Remember that financial wealth is only one measure of abundance. By manifesting their big ideas they opened the door to the universe of abundance. Thus they created the ability to materialize many of their desires. Which in the end have affected a multitude of people."

Let's look at some complex manifestations and inventions:

1. Man lands on the moon in 1969
   a. Thought – man should explore the moon;
   b. Intention – In 1961 President Kennedy says that he is committed to having man land on the

moon before the end of the decade;

    c. Attention – the U.S. government provides billions of dollars in financial budget and resources to do this;

    d. Manifest – Man steps on the moon in 1969.

2. Edison Invents the light bulb

    a. Thought – Edison must have had a thought such as 'I could invent some artificial light with electricity';

    b. Intention – I am going to make artificial light with electricity;

    c. Attention – tested 3000 different ways of producing artificial light. Each time he failed he crossed that method off the list until he succeeded;

    d. Manifest – Edison invented the light bulb.

"Uncle Jack, this is all starting to make some sense. I am working with you as you requested, but I have set goals such as having my own cottage in Pine Ridge one day. However, it hasn't happened," stated Tom.

"Tom that is a great thought and I wish you the best in making that thought manifest. It really is that simple; thought, intention, attention and manifestation. Bear in mind that complex or large goals can take time and perseverance. The challenge we face is that we prioritize our intentions and re-prioritize them on a dynamic basis. Scientists believe that we have about 60,000-70,000 thoughts a day and most of these thoughts are the same thoughts we had the day before. We have both positive and negative thoughts. So which thoughts would we like to focus on and which ones would we like to cancel out? That's easy to conceptualize but even more important to be aware of as we go about our random thinking in a day. Of these 60,000 – 70,000 thoughts we have in a day, there would be many intentions, including the big intentions we have in life. These intentions start to compete with one another and many times we sacrifice our big intentions for a quick fix. So the attention you put to your most important intentions must be a burning desire (attention) and it may need complete focus over a period of time for it to manifest. The last ingredient to the magic is that you need to set some type of time parameter for the goal to be materialized. What I mean by that is that you to set a target date that is realistic within your current circumstances."

"Manifestation relates directly to our lessons in quantum mechanics. Deepak Chopra stated that the world of materialism is dead. He said that everything is made up of energy and information and this is where our lesson about quantum mechanics comes into play with manifestation."

"What do you mean when you say the world of materialism is dead?" asks Tom.

"Remember the other night we discussed the fact that the smallest building blocks of the universe are not solid material objects at all. They are all made up of energy and information. In the Newtonian world, we didn't consider quantum mechanics, relativity or $E=mc^2$. Everything in those days was related to physical or material things. The world we live in today is very different and we need to embrace that thought. We live in a world of possibilities and it all starts at that quantum level."

"Let's do a quick review on atomic structures. The smallest collections of objects known to man in the early 20th century were atoms, which are the building blocks of the universe. These building blocks are carbon, hydrogen, and helium among others. However, as we discussed, these atoms in turn are made up of smaller bits with names such as electrons, protons, neutrons, quarks, and bozons. The only way that the components of an atom can be observed are by the trails they leave behind in massive particle accelerators. These atomic elements cannot be observed visually – the smallest observable structure is an atom."

"How do we know they are there?" asked Brian.

"We know they are there because of the results of Rutherford's and Bohr's work in the early 1900's. Now we have facilities called Particle Accelerators that can test for these elements. There are only a few of these particle accelerators located in the world, like the CERN facility in Switzerland. The method they use to look at these components that make up the universe is by smashing the minute particles. When they do this they are able to measure the components. Much like if you threw a pumpkin at a wall, it would break and you could observe the contents. However, the only proof of the existence of a quark is by measurement with very sophisticated devices. Even then, the evidence of these particles is displayed as a wave or a probability. There is never any observable material object, just a trail of its existence."

"The only difference between material objects such as gold, silver or wood is the arrangement of the energy and information. They all contain carbon elements, but in different quantities and arrangements. At the quantum level it is only energy and information, which is non-material."[1]

Uncle Jack pauses and there is a collective sigh from the mesmerized audience.

"It really is something when you start to put this together, isn't it? Although I find it all very exciting, it gets even more interesting as we continue," exclaimed Uncle Jack.

"We really have gone past the material Newtonian world that some of the ancient philosophers and scientists experienced and hypothesized. Keep in mind there were others who understood manifesting and the non-material world thousands of years ago, before we had as much evidence as we do now. For instance we have already unknowingly abandoned the material world in our daily lives, without even a second thought about it. Think of an example in everyday life that might prove this point, for instance, when you pick up a mobile phone and call home you just dial the number and don't think about the magic required to make it happen. How can we leverage this same magic for our own needs? You can't see your voice between the phones when you are talking and you can't feel it either. You can't even conceive of what type of form your voice is in as it is transmitted from one mobile phone to another. Yet your voice, which is simply manifested thought, can be carried around the globe. Mobile phones, portable phones, fax machines, television, radio and computers are all based on energy and information, not the material world."

"If devices such as mobile phones work on energy and information (non-material stuff) think of the power of your mind in being able to manifest your dreams. What else are we capable of? That is a fascinating question and it propels me further in my research."

"There are two conclusions that are important when we consider the theory that the smallest objects in the universe are made up of energy and information:

1. Real material objects that you can touch and feel, like a chair for instance are *real* objects in our world, but they are made up of nothing more than energy and information; non-material stuff.

2. The second and most important thing to remember when we discuss manifesting is that *thought* is also energy and information. So, *both a material object and a thought are the same thing; energy and information.* This is a subtle, yet very powerful point."[2]

"When you complete our time together you will have learned many lessons. Some will be lost in the archives of your brain waiting to be retrieved, and others will stick in the forefront of your mind for the rest of your life. This subtle point that I just discussed is one which I would ask you to remember. It is the reality of life. Everything is energy and information. They are the same thing."

With that, Uncle Jack paused and stoked the fire. The brilliant orange ashes sparkled into the night sky as they all stopped to ponder that thought.

"This is magic. A thought changes form into a material object when it is manifested. Energy doesn't disappear it just changes form as stated in the equation $E=mc^2$. This cycle continues forever. Thoughts are very powerful."

Almost as if a light bulb had popped on over his head, Tom says, "Uncle Jack, I think that I am beginning to understand what Einstein talked about even more clearly when we talk about manifestation. A material object turning into energy at the speed of light makes sense to me now because it's just like turning a thought into a material object."

"Tom, that was a great connection." exclaimed Uncle Jack.

"What is also important is to take the time to reflect on what you have achieved. Think of the things that you are doing now or recently accomplished and you will realize that you are living the life you wish to live. You are doing a lot of the things that you wish to do, like coming up to the cottage, doing activities with your family and friends. You are living your dreams today. So, if you want to change other things in your life, dream big and as the famous Nike ad extols – *just do it.*"

"Before we depart for the evening I want to leave you with a thought. There is one final ingredient to the recipe of manifestation and that is for you to focus your attention on the present or what I call the *now*. The present time is the only time you physically have a chance to make an impact. The future or the past are energies that are virtual images. You can picture these images that are stored in your mind like movies on a video tape or DVD. The present is the only real moment; it is the key to your future. So, don't struggle with the thought of how your current action achieves your goal at every moment in time. Focus on what you are doing at that moment in time. Put all of your attention and energy into your current action. You must surrender your intentions to the universe and focus on the *now*."

"However, what I do recommend for you is to review your

goals on a regular basis. Perhaps daily, weekly, monthly or annually so that the future is clear and then as Deepak Chopra states, *the universe will handle the details.*"

"Don't you agree we have had enough for this evening? My intention is to go off to bed." The others agreed and picked themselves up off the beach and shuffled up the stairs to the top of the cliff. They headed off to bed with visions of the future manifesting in their heads. As usual, Uncle Jack extinguished the remains of the glowing coals with water from the river.

Key messages or ideas to take away from this Chapter:

1. If material objects are made of energy and information, then what is thought? It is energy and information. They are the same thing.

2. Thoughts are powerful.

3. Everything that was ever manifested started as a thought and then transformed into a material object or occurrence.

4. Focus on what you want, not what you don't want. You will manifest your intentions, so be careful what you intend.

5. Leave the Newtonian worldview behind. There is much more than the physical world we see, feel and touch.

6. There is only one thing stopping you from achieving your goals and dreams; and that is you.

7. Manifestation is magic. We all do it.

8. You control your ability to manifest your destiny. You can change your dreams, dream bigger dreams, it is up to you.

9. What are the mechanics of manifestation?
    - Intention – set a goal
    - Attention – focus and take action
    - Manifest – achieve the goal

10. Focus your attention on the *now* and surrender your intentions to the universe.

7 *Obstacles or Opportunities in Disguise?*

Problems cannot be solved at the same level of
awareness that created them.
—Albert Einstein

There was a strong warm wind blowing across the river filling
the colorful sails of boaters and boarders. The largest sail-
boats were in full sail, tacking to and fro as they toured back
home. A few windsurfers were blasting over the white caps
harnessing nature's power as they sped across the river. Back
on the beach the occasional strong gusts of wind made havoc
of picnic plans by blowing plates and cups into the shallow
water at the river's edge. Umbrellas and picnic lunches had
to be tacked down to keep them in place. Eventually all beach
paraphernalia had to be secured or stowed away. Meanwhile
hoots and hollers could be heard from the surfers as they
dunked their heads backward in the water as their boards accelerated across the river in the strong gusts of wind. As the
day wore on, the wind died down and the surfers returned to
the beach and shared their exhilarating experiences of catching the strong winds, jumping over the waves and of crashing
into the water.

By evening the air was still and all was quiet. From the
cliff, high above the beach, Tom, Brian and Carolyn were
watching the sun as it began its descent behind the distant
tree line and hills in the west. The sun was a glowing red ball
that created a blanket of red and orange hues on the horizon

and when it finally disappeared its colorful blanket remained for just a moment as if it were a final calling card, a reminder of a beautiful evening.

Down on the beach, Uncle Jack and the youngest children were already roasting marshmallows and singing songs. The combination of wood smoke and voices carried up the cliff to the others who would eventually drift down to join the fun. As dusk turned into full darkness all that remained at the bonfire were Uncle Jack and his students.

"Tonight I want to talk more about manifesting but I want to talk about it from a different perspective. I want to focus on what happens when we encounter obstacles. It would seem that obstacles and challenges are lurking everywhere. In fact, obstacles are a component of manifesting. Just like taking action. However, the question is, what do you do about them?" asked Uncle Jack.

The group had an introspective look on their faces, perhaps not quite sure where he was leading them.

Uncle Jack continued, "Sometimes you vigorously attack obstacles and challenges. Sometimes you roll over and play dead; or, you just give up. Why? That's what I want to explore this evening. What motivates you to push forward when obstacles are in your way and yet other times you give up?"

"I think it has to do with confidence," said Tom.

"I agree with you Tom, but there is more to it than just that. What if you set a goal or intention, and then apply attention but you are not able to materialize your goal? Perhaps frustration sets in and you get discouraged. At this point, depending on how discouraged you feel, you either push forward or quit. Maybe negative self-talk causes this discouragement, or perhaps a friend or peer criticized you. Or perhaps the negative self-talk sets in before you even started and you never get past the starting line. Have you ever had this experience?"

They all nodded.

"Of course you have. It's part of the human experience. Confidence is part of it, frustration is another, but ultimately it is pain or pleasure that motivates you," said Uncle Jack.

"Pain or pleasure?" questioned Carolyn.

"Tonight's discussion won't be painful will it Uncle Jack?" queried Brian with a smug look on his face.

"No Brian, it will be a pleasure." countered Uncle Jack.

"Yes that's right. Pain and pleasure ultimately drive your desire to move forward. Sometimes the pleasure of achieving your goal will be enough to motivate you to go forward. Other times it won't matter how much pleasure you gain through this experience, you won't pursue that particular activity, until the pain you experience in *not* pursuing it becomes unbearable. Conversely, if the pain is well recognized you will act immediately. This pain/pleasure mechanism underlies everything you do," said Uncle Jack.

"Here are a couple of quick examples. People will avoid doing things because of fear or pain. For example, many people don't like public speaking so naturally they won't go looking for opportunities to speak. The fear of public speaking is too painful. There is the fear of humiliation that they might be unable to communicate their ideas, forget their words or perhaps be intimidated by the audience. You could describe this fear as a de-motivator. On the flip side, fear of poverty drives people to find and keep jobs. In this case fear is a motivator. Pain is driving people to action. Perhaps even to the point where public speaking is a must."

He continued, "Negative self-talk is your biggest personal challenge or obstacle. It's your fear talking to you and it can be very persuasive. It will convince you that the pain will be great and therefore you should forgo your actions and goals. This could be a good thing if there was impending danger to your life. For instance, imagine you are about to cross the street at a busy intersection, cars are zipping in and out of traffic and this sight causes you to second guess your plans as you pause to consider your safety. This fear is ingrained into our genetic make-up for good reason. It keeps us alive. However, let's imagine the pursuit you have is not one that will cost your life, it is the pursuit of a legitimate goal or dream. Fair enough?"

They nodded their heads.

"Negative self-talk is inherently limiting. You are essentially trying to convince yourself that you will fail. In the end you may succeed, however you are limiting your chances before you start. There are many phrases that we use to express negative self-talk. Some are expressed outwardly; others are expressed only in the mind. Here are some examples."

§ I can't do it.
§ It costs too much.
§ I don't know how to.
§ I don't have enough time.
§ I don't have enough help.
§ The economy is terrible.
§ There is too much uncertainty.
§ I don't have enough confidence.
§ I wouldn't know where to start.
§ No one is willing to help me.
§ I wouldn't want to bother anyone.
§ What would they think if I did that?
§ What if I stumbled and embarrassed myself?

"Do they sound familiar? Have you ever caught yourself using these statements?"

"I must admit that I've used many of these, but sometimes I push forward anyways. I can think of days at the snow park where I was afraid to try some new tricks," said Brian.

"What is the genesis of these negative comments and Brian's experience at the snow park?" asked Uncle Jack.

"I think he was afraid that he would hurt himself or perhaps embarrass himself in front of his friends," said Carolyn, in such a way as to preserve Brian's integrity.

"I agree with you Carolyn. If he truly believed that he could achieve the trick he would have been more confident to try it. Confidence comes from past experience or blind faith. In this case the pleasure of succeeding was overshadowed by the fear of not succeeding. However, on the flip side, if you knew you wouldn't hurt yourself and you didn't care about who saw you fail or succeed, you may have gone ahead and tried your trick," said Uncle Jack directing his comments to Brian.

"That's right, I agree with you. I hate to admit it, but it's also true that I don't want to look like a rookie falling in front of my friends and filling my goggles with snow. That would be painful," answered Brian.

"How does a snowboarder learn new tricks? How did you do your 360^0 jumps the first time?" asked Uncle Jack.

"I just got to a point where I convinced myself to try it. I saw other snowboarders I knew trying 360's, 540's and 720's, so I thought I might as well give it a try. I had a vision of myself doing the rotation. Was I scared? Absolutely. It took me a number of times to do it. I had a number of spills without injuring myself and eventually I was able to do a full rotation and land. I have had to practice my boarding tricks over and over until I was able to consistently complete them in competitions. I think you'll find it interesting that I still get a lump in my throat as I start my descent for the jump. That's why I do it. I love the thrill and not everyone has the guts to do it. I guess in my case, overcoming my fear and the successful vision motivate me." exclaimed Brian.

"Brian you should be proud of your snowboarding capabilities. They didn't come to you by accident. You are right, it takes practice but more importantly it takes confidence in your abilities. This is no different from achieving any goal. The key point to understand here is that despite your fears you tried it. And even when you didn't succeed the first few times, you kept on trying until you succeeded. Failure just slowed you down and then you kept practicing until you perfected it. This plays very well into our lesson tonight. The pleasure of success was driving you, not your fears."

Uncle Jack stopped and gave Brian a high five and everyone chuckled.

He continued, "This is something we pose to ourselves in every encounter we face; is it possible for us to take on this challenge? Anytime you are faced with something new there is a challenge or obstacle in your way. When this occurs you are facing the unknown, as Brian did. As we learned in our lessons on quantum mechanics, there are a number of potential outcomes. There are no pre-defined outcomes or a pre-destined answer. You have to be comfortable with facing the unknown and stepping into it. The unknown is the environment of creation and all creation happens with uncertainty."

He paused as his students contemplated that thought.

"So instead of saying or thinking "*I can't*", reframe that thought with "*it's possible.*" When you hear yourself thinking negatively it is your opportunity to change that thought. Awareness is critical. Ask a trusted friend/peer at work or school etc. to let you know when you are approaching a problem from a negative perspective versus an *anything is possible* perspective. By becoming aware of this you will leverage laws of quantum mechanics as in Heisenberg's uncertainty principle. The mere act of observing yourself in this state will bring awareness and will help you change your perspective."

"The simplest thing you can say to yourself or to someone with whom you have reached an impasse is to say – *I realize we have a misunderstanding, however, I trust we will be able to work it out.* That is a very powerful statement. It acknowledges the obstacle, respects the other person's perspective, and reframes the situation completely so that you can jointly seek out an alternative solution. This simple concept invokes a magical process that helps people overcome disagreements or challenges in working together. Again, awareness is the key."

fear

Zone of Comfort

safety/pleasure

fear

fear

fear

"We all live in a zone of comfort. Typically in order to move out of that zone we are uncomfortable; perhaps even fearful. There is an uncomfortable feeling when one stretches his or her zone of comfort. This is a universal feeling. Professional athletes, actors, politicians and business people all feel similar feelings as they step into the unknown. They get butterflies, panic feelings, anxiety etc."

"Your body feels the fight or flight reaction to stressful events such as job interviews, exams, public speaking, dealing with relationship issues or life threatening circumstances. This reaction has been locked in our genetic code since the early history of humanity; it is our natural reaction to fear. If we are in danger the body automatically creates a chemical reaction, which provides us with the tools to react to danger. Increased adrenaline increases our heart rates so that we can sustain the effort to run away from what caused the fear. Your success in overcoming these feelings is very important. While the fight or flight reaction is encoded in our genes, we no longer have to react to a job interview like we did in prehistoric times when predators were stalking us. By overcoming this fear or harnessing it and understanding it, you can grow and expand your comfort zone."

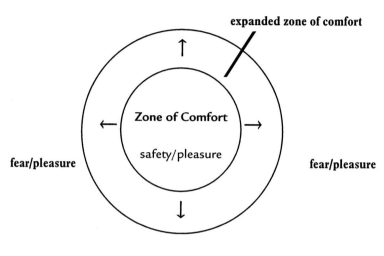

"Again, motivation is interesting. At times we feel fear in our current situation and conversely pleasure by growing our comfort zone. Other times we are happy with the status quo. Remember this, status quo is stagnation. There isn't any growth within stagnation and therefore no movement towards your goals and dreams."

Tom jumped in, "I never realized that this feeling was as universal as you described. Sometimes you feel like you are the only one that has these feelings, however, it seems that everyone feels the same way when they step out of their comfort zone. It makes sense."

"When people are out of their comfort zone they can act very uncharacteristically. The feeling of fear can create many different reactions that may vary from being completely withdrawn or lashing out in anger. For instance, a coworker or one of your friends could be reacting with anger and lashing out because they are fearful and out of their element," said Uncle Jack.

"When you recognize this in other people you can disarm them by using a few simple questions or statements such as:

1) Is everything okay?

2) Looks like you are having a tough day.

3) Is there anything I can do to help you?

This disarms a fearful person because they immediately become aware of themselves and have to look within. I would also advise that you show genuine empathy toward them. Try it. Again, awareness is the key to making change."

"I've noticed some very erratic behaviors in people at work, generally when we have been under pressure. There are certain people who always seem to lash out and react very negatively," said Tom.

"Whenever you work in large groups there are some people who can be very counterproductive. Their energy can be very negative and can in fact bring a whole group down. It's pretty obvious that when you are working in a team environment that positive energy is a must. Separate yourself from negative energy and be aware of your own attitude. By becoming aware of your attitude you can alert your mind to your own negative self-talk. It is powerful. Challenge yourself to *"think outside the box"* and whatever you do leave the negative self talk in the box."

"I don't mind when people at work take a devil's advocate perspective on a potential solution to a problem. I think it's healthy to look at the solution from many perspectives. I just can't stand when people jump in and throw roadblocks at you when you're trying to brainstorm a solution. It drives me nuts. It takes the wind right out of my sails. The funny thing is that I don't even think that they realize how destructive their attitude is. You just don't want a person with this type of attitude around when you are trying to solve a problem," exclaimed Tom.

"You are describing what I would call a pessimistic person," said Carolyn.

"Yes, attitude is altitude. Surround yourself with positive uplifting people and you will draw upon their energy. This is critically important when working with people. Positive energy is what you need, not negative. My recommendation is to fire negative people from your lives, whether they are work associates, friends or even family, but that's for our discussion tomorrow night. Let's get back to our challenge at hand," said Uncle Jack.

"Whatever it is that you want to pursue you must give it time and be persistent. As we discussed last night this is what I mean by saying you must give your intention, *attention*. Having said that, not everyone has the opportunity to be the President of the United States or to be Neil Armstrong the first astronaut to land on the moon. The reality is that you must work diligently towards your goal, stepping out of your comfort zone. As soon as you attempt to do this an array of possibilities will open up to you that did not exist before. These possibilities appear just because you have taken action. All you may see today is one or two potential paths, but as you continue, you may realize that you could pursue a multitude of possibilities. In fact that's what happened to one of the presidents of the United States, Ronald Regan. I don't suppose Ronald Regan ever thought that he wanted to be or even thought he could be the President of the United States during his early movie star days, yet he eventually quit acting and became the Governor of California and parleyed that position into becoming one of the most popular Presidents in U.S. history."

"If I lived to be a hundred years old there could only be 25 presidents in my lifetime, assuming a 4 year presidential term," computed Carolyn.

"That is correct Carolyn. Twenty-five presidents is a very small number considering a population of approximately 300 million. That speaks exactly to my point earlier about how other opportunities will present themselves. This is not a bad thing," replied Uncle Jack.

"However, if you don't achieve your goals, frustration and discouragement may set in. Frustration, discouragement and anxiety are signs that you need to take into consideration," stated Uncle Jack.

"What kind of signs are these?" asked Brian.

"They are signs that are telling you that around the next corner is a potential opportunity and that some type of change is required. This change could be something small or could be a change to your life strategy. If you are not consciously aware of these signs you'll miss them. You may mope or cry about the fact that you can't accomplish your goal. This behavior is easy. Anyone can mope and cry. As you get frustrated and discouraged all you really need to do is bear down and think of alternative approaches and harness that energy to propel you forward. You need to step back a bit and look at it from another perspective. Sleep on it, meditate, do some other task for a while. Einstein once said, 'you will never solve a problem by being in the same mindset that you were in when you created the problem'. And as we all know he was a smart guy who tackled many challenges in developing and proving his theories."

"We have certainly come to know that fact Uncle Jack," said Carolyn.

"When you perceive that you are frustrated, discouraged and think you have failed, ask yourself what have you learned from this experience? Using a military analogy, you may not have won the battle but the war is far from lost. This lesson is an important piece that will define your future. Just as when Edison invented the light bulb, it took him 3000 attempts to develop a light bulb before he was successful. He was fond of saying that he found 2999 ways it didn't work. With each battle lost, he learned something and applied this knowledge and he continued to wage the war until it was won. In fact, many discoveries are made by accident, for instance 3M's Post-it Notes. 3M scientists were researching new glues when they stumbled upon the idea for Post-it Notes. Failure is not the end of the game. It is a new beginning."

"Whether you are at work, school or play, what you experience today is a building block for your future. Success or failure provides an opportunity to examine what worked and what didn't. If you failed to do what your customer expected or what your manager requested or what your teacher or professor expected, it is your opportunity to ask questions and get their perspective on how to make changes to your approach. The next time you are up against this situation, you will have the wisdom through experience to overcome that particular obstacle or challenge. This is why work experience is of such great value to employers and is required information on a job application or resume. People without experience are often discriminated against because they don't have the perceived value that someone with years of experience does."

"I've always hated hearing that I didn't have the experience to do a job," said Tom.

"I hear you Tom. On the converse side I don't want to ignore the burning desire that people have to overcome a lack of experience. Pure desire is the other necessary ingredient. Organizations also like to hire inexperienced people who are willing to learn and are prepared to do what it takes to succeed. This burning desire coupled with experience is an incredible combination for success," said Uncle Jack.

Uncle Jack stoked the fire and put in some more wood, which quickly illuminated the area when it burst into flames.

As he sat down he continued, "Bigger goals have bigger challenges. Smaller goals have smaller challenges. It's all relative. Sometimes you are able to remove the blocks all by yourself, other times you can't even see that you are blocked."

Brian jumps in, "like the expression, you can't see the forest for the trees."

"Yes, you're right. It may seem evident that you are not achieving your goal, but you are not clear as to why you are slipping sideways relative to your plan. Sadly your attention falls elsewhere and your manifestation plans fall into neutral."

"For instance, you can't expect a child to know how to do long division in mathematics when they are in pre-school, unless they are exceptionally gifted. You can say that a pre-school child who can't do long division is an extreme case of being blocked or challenged. However, it is the same case for adults or teens when they come up against a challenge that looks insurmountable. The child needs a few more years of maturing and some basic math understanding before they can do a long division question. Eventually all children can do long division unless they have severe learning disabilities. They just need the skills that education provides to get them around the block or challenge long division presents. You can't expect the person working in the mailroom to be promoted to the position of the president of a billion dollar multinational company unless he had the training that job requires. The skills between these roles are vastly different. It is, however, absolutely possible for this same person to become the president of a multinational company but they will need time and energy to learn the lessons required of that role."

"So, let's get back to obstacles and challenges and concentrate on your goals and dreams. The question remains: what is stopping you from achieving your goals and dreams? Is there something side tracking you?"

"That is a good question," said Carolyn.

"Well let's face it. It's easy to do the easy things. It is less painful. For instance, at home it's easy to fritter away your time living in your comfort zone watching television instead of going out to fulfill your dreams," pronounced Uncle Jack.

"Uncle Jack, television can be a great learning tool and it is often very relaxing and entertaining, however, I agree it shouldn't become your life," said Tom.

"Tom you are absolutely right, television is a great source of information and entertainment. However, it is a pastime that I think is over used in many households and becomes the default activity for many instead of living life to its fullest. Therefore let me use the example of television for the next little while, as it competes for our time and energy and thus reduces the ability to achieve goals and dreams," said Uncle Jack.

"In many homes the television is in the most popular room of the house and it's constantly on whether someone is paying immediate attention to it or not. I agree it is a great learning tool and provides relaxing entertainment, however, it is very passive. In other words, you are actually watching other people live their lives and execute their goals whether it is in a movie or a sporting event. Why not do the specific activities yourself? It seems to me that some of us are living a voyeuristic life," claimed Uncle Jack.

"I see what you mean. You are hoping we will participate versus watch. In fact, that's what we do here in Pine Ridge," said Carolyn.

"If you have goals and dreams and you wonder why you're not achieving them, ask yourself how much time you spend doing activities such as lying on the couch watching television? Where are you applying your attention? If you are spending a great deal of time watching television, you are manifesting a different dream. You have become a very good television watcher. The reality is that you have become very good at being a couch potato."

He paused, knowing he was going to get a good belly laugh from his students.

"How much time are you wasting that could have been converted into action, which would thrust you towards your goals and dreams. How much closer will you get to your goal, by taking a course, meditating, investing in your relationships, researching information on your own or actually doing what you are passionate about? Your biggest complaint could be the lack of time you have to accomplish your goals, but perhaps the crux of the problem lies in how you are spending it. Ask yourself, whose life are you living: CBS's, NBC's, CBC's or your own?" prodded Uncle Jack.

"It's easy to do the easy things like falling onto the couch and watching the tube after work or school. During the day it's easier to go through the motions. It is always more difficult to do what is passionately excellent, whether you are at work or school. It costs energy and time. It forces you to get out of your comfort zone. It's risky, it's scary, it distinguishes you, and it centers you out in a crowd. Does this sound familiar?" asked Uncle Jack.

"So the question is: what's stopping you from achieving excellence? Sadly the answer is that you are," exclaimed Uncle Jack.

The three students knew that Uncle Jack was right on the mark with his message.

"Let me get away from television for a moment and focus on what motivates you to either take action or refrain from action. As I stated earlier, everything comes down to personal pain versus pleasure. You may perceive that extra work will cause you too much pain in spite of the pleasure gained from accomplishing a goal. Perhaps becoming the center of attention may be beyond your current comfort zone. If you want to step up your game you must take the leap out of the comfort zone. Otherwise you will remain in the twilight zone, never actualizing your dreams. It's your *choice*," said Uncle Jack.

"Uncle Jack, does it always come down to pain versus pleasure?" asked Carolyn.

"Yes it does, but there are some twists associated to behavior. Let me give you an example that most of us can relate to: Chuck thinks he needs to lose 30 lbs to get back to his trim self. He may think he really wants to, but his behavior says something else. Everyday when he wakes up he promises that he is going to stop eating ice cream and exercise more often but by the end of the day he is chowing down on a bowl of Ben and Jerry's Chunky Monkey while watching the Weakest Link. What is happening is that the pain of cutting down his food intake is more painful to him than the pleasure of the vision of a svelte and healthy Chuck."

"From the outside Chuck's ordeal looks very simple, but yet he can't do it. Although it is his decision, he is still unable to discipline himself. The pain of giving up food is greater than the pleasure of a trimmer self. He is currently in his zone of comfort, 30 pounds overweight, snacking when he feels like it and eating whatever he wants. Despite how negative he feels about himself, the comfort that food provides outweighs his vision for himself. When will he lose weight? This is up to Chuck, no one else can make this decision for him. Perhaps he will never fulfill this vision and he will die overweight."

"How does a person like this motivate themselves to change," asked Carolyn.

"Perhaps his health will be compromised and then the pain/pleasure equation will change. Then in order to resolve his health issue, losing weight will become a life and death issue; suddenly the pain of death and the pleasure of life are compared. The likelihood is that he will go on a program to lose weight. Unfortunately it is these life or death events that often drive people into action to lose weight, repair broken relationships or to see life as a precious gift. Fear is an incredible motivator. Unfortunately it is not a motivator that anyone should choose, because many times fear is the last option and by that time much damage has been done. Often fear presents itself as the last resort. In not taking any action the consequences become significant such as: death, bankruptcy, divorce, illness etc. You are now basically running in fear for your life. If you stop, you are dead. How does that image grab you?"

"I don't like it one bit," said Tom.

"The NASA space shuttle computers make millions of adjustments to its pitch and yaw as it glides in for its landing.

In fact thousands of these corrections occur every second. These corrections may not be apparent to observers, however, it makes the landing routine and smooth. This same principle applies to Chuck. By making better decisions every day, Chuck will improve his chances of a long and healthy life. It may not be thousands of corrections per minute, but it could be 5 or 10 better choices in a day that affect him. Including decisions about portion sizes and the kinds of foods he eats," said Uncle Jack.

Tom smiled and said, "It is amazing to think that as few as 5 or 10 decisions a day could make such great changes."

Uncle Jack added, "They could be as simple as:

1. Reading an inspiring book for 30 minutes in the morning or meditating.

2. Eating a healthy breakfast.

3. Eating a healthy lunch.

4. Eating a healthy dinner.

5. Exercising in the evening – walking, gardening, running, golf etc.

6. Drinking water all day instead of caffeine and sugar drinks.

7. Spending time with loved ones."

"The reality is that Chuck's weight issue is a bigger issue than just his diet. It is related to his lifestyle and his decision-making ability – specifically his discipline. Diet is one part, exercise is another, but his behavior is the key. Chuck, like everyone else, knows that diet and exercise are the keys to change but he can't change. Overeating is a negative behavior that can't be eliminated, only replaced with a more positive behavior. Food fills a void just like television does. Food and watching television are only symptoms of this pattern. Chuck has a "*so what?*" attitude. Chuck does not care if he wastes time in front of his television or he eats foods in excess. If your life is framed by a 'so what' attitude, it's time for re-framing. A 'so what' attitude invades everything. There is no thought to stepping out of the zone of comfort. It is too painful for Chuck. This is not to say that Chuck will not achieve goals and dreams but he really is limiting himself. His capability and his attitude are wrapped up in pacifying himself with food or television. Then he shrouds himself in negative self-talk which in turn convinces him that he is not worthy."

"Well that's a pretty bleak picture," said Tom.

"This example is common. The behaviors are different for everyone but the concept is the same. The negative behavior fills a void. The void could be due to the loss of a loved one, an unfulfilling job or relationship, loneliness, lack of self-respect etc. So what do you do with a negative behavior? You must find positive behaviors to replace the negative ones. It's that simple."

"When you see a person who has started to manage their weight, you see a big change in their attitude and behavior. You see confidence in many other areas of their lives. They eat a high performance diet and they participate in exercise. They reenergize all facets of their life, stepping out of the malaise. You see it in their relationships, their work and their studies. They are out of that former zone of comfort. They are excited about their life because they are taking action. New possibilities exist. They are making better decisions about food, exercise, and their general day-to-day activities. They are controlling their lives and their destiny. They see the pleasure in making better decisions. They are not running for their lives, they are living it', said Uncle Jack.

"Once you can do this you are a successful 'life' manager, finding discipline as a small pleasure, a small victory on the way to self-actualization. Small corrections along the way fine-tune your life, just as the shuttle corrects itself on landing," explained Uncle Jack.

"The shuttle really provides a good analogy," said Tom.

"If whatever you try doesn't work, you have learned something. It will either help you overcome the challenge with that specific activity or as the expression states: *as one door closes many more open*," said Uncle Jack.

"That's a great expression. One door closes and many more open. It is so true," said Tom.

"That's how manifestation works. You see a block or a challenge and you figure out how to get around it, over it, under it or through it. If you don't, you remain status quo, but that won't help you get to your dreams or goals. The point is that the power is at hand and is within everyone: but by doing the easy things you are never exercising the abilities you have. Some people discover their abilities much later in life. Why not get started now," said Uncle Jack.

"How?" asked Brian.

"Just do it, whatever it is, you will figure it out." He paused for a moment to think about how to answer that question. "I think the key to growth lies in your personal education. It doesn't necessarily have to be formal education; it's practical experience that you need. Your level of knowledge and ability will grow when you challenge yourself. I've always said, *you don't know what you don't know*. When you are at the edge of your ability, step up and experience the next level. You may fail or you may succeed, it is unknown until you try. Whatever result, the benefit is that you will grow. Just like when a baby is learning to walk. Take the risk. Step into the unknown and realize this is your path to grow and this is true at any age," exclaimed Uncle Jack.

7.1 *Universal Alignment*

When you want something, all the universe
conspires in helping you achieve it.
—Paulo Coelho

Insanity: doing the same thing over and over again
and expecting different results.
—Albert Einstein

"As we have discussed, thought is energy. It manifests itself into action or a material object or service when action has been taken. Action then is the catalyst to physically manifest any idea. It's this simple. First you need an *intention* to do something or make something and this needs to be followed up with *attention* to ensure it happens. For instance when we plant seeds in the soil in the spring, we must water, fertilize and weed the surrounding area and eventually the plant begins to grow. If there isn't enough attention the result may not be very successful."

"Actions are all related to thoughts. Some actions are subconscious or automatic like breathing or the need to eat or drink. Your mind/body manages these processes on its own without any conscious intervention."

"Other activities become automatic or sub-consciously controlled. Consider walking for instance, you don't have to think about every step you take although you might occasionally think of your speed or direction; typically you don't have to think about the mechanics of lifting your leg and then moving it forward and putting it down and then lifting the other leg etc. You just do it."

Brian starts imitating a baby learning to walk and has everyone rolling around in laughter.

"As a child you have to learn the basics to be able to survive and grow. A baby needs to learn how to communicate when it needs attention or food. It learns to communicate through crying. Crying develops into language. To move from one place to another the baby learns to roll, crawl and eventually walk," said Uncle Jack.

"As an adult we don't have to worry about how to walk, talk or to manage ourselves. We have learned these skills and they have become automatic or subconscious."

"What about when we take action and hit significant blocks Uncle Jack?" asked Tom.

"Yes, sometimes it is difficult to know where to start or how to take action to overcome some obstacles whether they be big or small. What do you do? Turtle? Picture the head of a turtle hiding in its shell until it's safe to come out. *Turtleing* may be helpful for a short period of time to collect your thoughts and determine your next action, but it is not a good long-term strategy. As you turtle, time is wasting. Don't let frustration and negativity prevail, persevere and direct your future. It's okay to go at a turtle's pace as long as you are moving forward," said Uncle Jack.

"Yes, I've done a lot of turtleing in my life," said Tom.

"We all have many goals in our lives. Some are very, very big. Others are simple, yet they remain as thoughts because no action was taken. Thoughts and ideas are swirling around in your mind continually. As we have discussed you have as many as 60,000+ thoughts per day. Some of the thoughts are the same thoughts you had yesterday, last week, last year etc. Some are very simple yet you devote a tremendous amount of energy ruminating about them and have many internal discussions that go on and on and yet you don't take action. For example how many times have you thought about calling someone and then procrastinate for days on end until you finally lift up the phone and make the call. Or, how often to you think about the benefits of going for a walk or a run but come up with excuses. When will you talk to your boss regarding a new job or about improved working conditions? Perhaps you haven't experienced these, but I'm positive you can think of a similar internal dialogue," said Uncle Jack.

"When the thought comes to your mind, exercise it or plan to do it and DO IT. You will free up a tremendous amount of time and energy that goes into thinking about it. Stop playing the game you play with yourself, such as, I'll do it tomorrow when I'm feeling more energetic, or when I have more time. The reality is there is no better time than *now*. If not now, you must prioritize it and schedule it. The key is to take action. If all you ever did were to have ideas and thoughts running in your mind and not take action, you would be living a virtual life. It would be all about 'what if I did it' or 'what if I didn't do it'. Need I say anymore? I think you understand," stated Uncle Jack.

They all nodded their heads.

"Many of the goals that we chase are simply symbols, whether they are love, money, education, homes, cars etc. The symbols represent the energy invested to reach a target, to overcome many challenges along the way. The process to reach these symbols is critical. It is in fact the most important part. For instance, the benefits of running a marathon, while inspiring, doesn't measure to the lessons and the benefits gained in the training for the event. For many people it is life altering. They become more aware of their health, their bodies and the commitment required to reach a significant physical goal. And again, whether or not they completed the event, they have learned a great lesson and received the benefits through their training regime."

"Let's take a deeper look at what it means to reach a symbol with another athletic example. A high jumper needs to learn specific techniques to compete at international levels.

They need to build up specific muscle groups and learn to visualize. At the Olympic levels they train hours a day in order to focus on their goal. Each jump takes seconds to accomplish, however, the physical and mental dedication to be a top high jumper takes years of practice. The obstacle in front of the high jumper is the high jump bar. It is a symbol. It merely represents years of practice in overcoming many obstacles on the road to getting to their current position."

"Keep in mind that some of the biggest obstacles are as follows: having the *confidence* to start in the first place, the *discipline* to practice and having the *perseverance* to continue. Gradually increasing the height of the bar, despite reaching plateaus at various points. *Conquering* the doubts, o*vercoming* criticism, e*liminating* the fear of competition and *maintaining* a healthy lifestyle."

"As they chase the symbol, the obstacle, the next great height, they are working on their discipline, their ability to manage stress, strive for goals and all the while they are overcoming obstacles and challenges," said Uncle Jack.

"Success or failure comes with a price. There is emotion locked into it. Disappointment, frustration, shock and sadness are a few words that describe how people feel when they meet with failure. Success takes tremendous energy as well. Practicing, training, stepping out there and getting out of your comfort zone all comes with a price. There is no free ticket. You must step out of your comfort zone and take action. Your comfort zone will expand as you do this and you will grow."

"You are no different in the pursuit of your dreams and goals. Your goals represent symbols that are important in your life. In pursuing your goals you learn to conquer obstacles that appear along the way. The pursuit is about growing and expanding your comfort zone. Facing fear and living in that unknown space that shapes your life. It doesn't get much more exciting than that as you commit to achieve your goals. Despite what anyone else feels or says, you are committed to this path," said Uncle Jack.

"Fear can be debilitating for some people. So fearful in fact that they are afraid to leave their homes. Those who suffer from this are agoraphobics. This is an extreme case of fear. Another common fear is public speaking, even thinking about it can make hearts pound."

"There are wide ranges of fears that prevent people from taking action. To overcome these fears, education needs to take place. You need to learn that whatever you are afraid of isn't the end of the world. I realize that overcoming fear can be a tough proposition for some and easier for others. Let's look at how some people can be blocked and not even realize it. Some people actually take on their fear and live their lives around this fear never wanting to tackle it. They have closely assimilated the fear as part of themselves."

"I'm not sure if I understand what you are saying", said Tom.

"For instance, a person with the fear of driving a car cannot disassociate themselves from this fear. They have essentially become comfortable living with this fear or disability. If they need transportation they can either take public transportation or ask a family member or a friend for a lift. The fear of driving becomes part of who they are and is accepted internally. This person can still be valuable to society, however, they have accepted a limiting belief. They may never want to learn to drive because they are too afraid. In this case the fear of *change* is more powerful than living with their current fear," said Uncle Jack.

"I think I understand what you mean. This could happen with any limiting belief such as with a person who feels they are overweight. They are afraid to lose weight because of the fear of change. They realize they would have to change the way they live and that is frightening. Therefore being overweight becomes acceptable," said Tom.

"Thanks Tom, you've got it," said Uncle Jack.

"I want to tell you a quick story about overcoming fear, but let me tell you the ending of the story first. In 2001 Barbara

Corcoran sold her real-estate brokerage firm The Corcoran Group in New York. She personally pocketed more than $40 million after selling her company, which was generating $2 billion in revenue. She started her own firm in 1973 at the age of 22 with $1,000 that she borrowed from her boyfriend. She had recently quit her previous job as a receptionist, prior to that she was a waitress in a diner. She had no other business experience and yet Barbara was exceptionally successful with her business the Corcoran Group."[1]

"The first time Barbara had to speak to a large group of professionals she panicked and left the podium just after starting to tell an opening joke. She had forgotten the punch line. As we have discussed, one of the biggest fears people have is public speaking. Barbara never gave up because she knew she had to conquer this fear in order to manifest her vision of the Corcoran Group. This vision provided her with enough pleasure to overcome the pain of public speaking and she was able to conquer her fear."[2]

"The pleasure of reaching your goals can drive your success, however, success or failure comes with a price. So how do you handle all of the inputs you receive in a day whether they come from your boss, customers, prospects, teachers, family, peers, friends and also deal with your own swirling thoughts? It can be overwhelming at times. Where do you disconnect to be able to think creatively and put perspective into place? When you are afraid or approaching a stressful encounter you do feel anxiety, stress or perhaps the fight or flight response? We all have felt these feelings at times. How do we overcome these feelings in a healthy manner?"

"Yes, I do want to know this. Please go on," stated Carolyn.

"Here is a thought you may never have considered. The body lives in the past while the mind is in the future. The illnesses, aches and pains you experience now are the result of your human experiences up to the current point. The body is the last place issues manifest themselves. The simplest example is how your body aches after strenuous physical exertion. A more complex example is when people have worked exceptionally hard without any breaks for weeks on end, or perhaps a loved one has died which pushes the person's ability to handle stress over the threshold of capacity, people get sick, have nervous breakdowns etc. It has been proven that many emotions create physical symptoms in the body. Western medicine deals with the symptom and doctors prescribe pain pills, anti-depressants without dealing with the underlying issue. Eastern medicine is an interesting philosophy as it treats the mind and the body together. Tai Chi and Yoga were developed in China and India respectively. They treat the body and the mind holistically. These practices all help release the emotions that are locked in the body and mind. Both Tai Chi and Yoga focus on being present minded."

"How does this help?" asked Carolyn.

"When do you ever free up the mind from your daily worries and connect with the silence? Most people are so busy they never do. They never take the time to be quiet and be within their own mind and body because they are busy participating in activities such as work, school, social activities, telephone/internet conversations or reading. The only time when they really disconnect is when they sleep. It is not enough in our hectic lives. When things get really crazy and our daily schedules are a complete blur even sleep can be compromised. Insomnia is a major issue with a large percentage of the North American population."

"Going within through meditative practice is critical for peace of mind. It is restorative, rejuvenating and it can also be used to focus on future vision. There are many books and programs that will help you develop this ability. I would challenge you to pick up any books on meditation or these ancient arts. This meditative practice also relates to fitness and hobbies where your mind is free from the typical demands of the daily grind. It provides an opportunity for creativeness. Many long distance runners describe a Zen-like feeling when they run. They are sometimes completely disconnected from their normal thinking patterns and from the physical running activity itself. So Carolyn, to answer your question, this is my first recommendation, find a quiet space each day and connect to the silence within yourself."

"Being present minded allows one not to dwell on the past or concern themselves with the future. The focus is on the NOW – the present moment. Freeing the mind and the body from the swirling thoughts. It links us to a higher level of consciousness. By disconnecting from the busy activities, the solutions and creative thoughts appear."

"As a society we fall into this trap of doing *busy things*. Many poor performers in the work world or at school spend a lot of time doing busy work instead of thinking at a higher level and contemplating bigger ideas."

"Not that I am a poor performer, but sometimes I end up tidying my apartment at school versus just focusing on studying for my exams," Carolyn added sheepishly.

Tom jumps in, "Or how I occasionally run around the office from meeting to meeting and I can't seem to grab on to the solutions for complex problems. Conversely, I can't tell you how many times I have come up with an answer to a business problem when I was out for a run in that *Zen-like* trance you were talking about."

"You've got it, when this happens you are connecting to the infinite intelligence by letting go of your busy thoughts. That is what meditation is all about."

"What if we can't get past big problems by ourselves?" asked Carolyn.

"That is a good question and I'm glad you brought this up. Sometimes we need help. If you had a physical problem like a broken arm or an infected tooth what would you do? You would seek help from a doctor or a dentist. There are professionals out there whose job it is to help you overcome these issues. If the problem is related to a fear or behavior that you are struggling to deal with, then you should seek professional help."

"Let me say that I believe the power is within you to overcome obstacles on your own. When you are really stuck it may be necessary to get someone to help you. There are people that can provide expertise in teaching you skills that you may need to help you uncover blocks and to help you with putting certain actions in place. Consider the following:"

§ Coach
§ Teacher, Instructor
§ Counselor
§ Psychologist
§ Trainer
§ Executive coach
§ Spiritual leader

"I don't want to dwell on personal issues. You know what they are. There are countless fears and behavioral issues from agoraphobia to alcoholism etc. They are inhibitors to your success and the fulfillment of your life. They are negative behaviors that stand in the way of your success. These behaviors need to be replaced with positive behaviors," said Uncle Jack.

"Don't rely on your family and friends to be experts. They can help but they only have so much expertise. Your friends and family do not have the tools to help you work through issues unless they are specifically qualified. You wouldn't want your friend to do brain surgery on you unless they were a brain surgeon would you?" asked Uncle Jack.

"I get your point," said Tom.

"Some issues and blocks are deep seated. Over the years you have been programmed within your environment. This programming comes from your family, friends, your workplace and through your formal education. Professionals such as psychologists, psychiatrists, counselors, teachers, doctors and coaches have been trained in their field to help people heal deep seated blocks. They see patterns within people and deal with similar issues on a daily basis. If you think you have deep seated fears or blocks, seek a professional for help. If you bury these issues they will surface and compromise your health. This is a guarantee. By burying these thoughts we will eventually manifest some type of physical illness. There is a mountain of evidence that supports this. The mind and the body are inextricably connected. For instance, anger may eventually manifest itself in heart related issues," said Uncle Jack.

"To supplement professional help you may consider the plethora of self-help tools that are available such as the Internet, books from the bookstore or the library, videos, audiotapes and CD/DVD's. There is a world of expertise out there waiting for you. When the student is ready, the teacher will appear, in whatever form you wish," said Uncle Jack.

"Spend the time, investing in yourself. Learn all you can as awareness is an unbelievable tool. Awareness changes the possibilities all on its own. We learned this through quantum mechanics. By becoming aware we are changing the playing field just as Heisenberg described in his uncertainty principle. The mere observation changes the outcome. Like shining a photon of light effects the position of the particle."

"I knew there was a link to quantum mechanics somewhere here," said Carolyn.

"Take a moment and imagine a couple of possible scenarios for your future. Staying status quo or making changes that you continue to tweak as you grow. What are the possibilities? Now imagine yourself in your 80's and 90's telling your grandchildren what it was you did in your life and the lessons that you learned. How important it was to develop friendships, to love each other and to passionately pursue your goals and dreams. Now visualize your gravestone. What does your epitaph say? What do you wish it to say? He or she was a faithful employee, who worked 9 to 5 and on weekends and nights or do you wish it to say that he/she loved life and lived it to its fullest pursing their goals and dreams improving life for their family and many other people along the way. The point is that you get to choose your life. You get to choose what is important to you. Be proud of yourself. Lead your life. Rejoice in it. Step out of your zone of comfort," exclaimed Uncle Jack.

"That is a great visual," said Tom.

"Don't limit yourself. Don't put fences around what you

think you are. The world is full of magic. If objects expand infinitely as they approach the speed of light and time literally slows down as Einstein describes, ask yourself the following question: Is the world as fixed as you think? Or, is the mind perhaps more powerful than you think? Regenerate yourself by spending time with your loved ones, meditating, enjoying hobbies, reading a great book, writing a letter, taking risks. Do what you really want to do. Take the challenge."

"Before we head off for the evening, I want to go back to our discussion about manifesting. We have spent a great deal of our energy discussing obstacles and challenges, which are inextricably linked to manifestation. Perhaps one of the most important thoughts to consider is that once there is total alignment within oneself and your goals, the universe will conspire to deliver to you at an unconscious level. Life is merely a reflection of itself."

"Excuse me?" said Brian with a puzzled look on his face.

"What I am expressing here is that when you are in complete alignment or congruency with your goals and dreams, they will manifest with little effort. For instance, when you share your love, it comes back to you like a reflection in a mirror. What you project is what you receive. It is it surprising to you that when you give unconditional love that you receive love in return?"

Tom sighed, "I see what you mean."

"Alignment or congruency comes into play with all of your goals and your life. When your actions come with a *clear vision, passion, desire* and *commitment* you are in *alignment*. Think of golfers in the USPGA professional tour. They all have clubs and golf balls, they all practice their games and have learned to overcome challenges and obstacles in tournaments, specific golf holes, and in their personal lives. Having said that, did you know that the difference between the top 7 golfers in the USPGA in 2003 was less than a one stroke average per year and only two strokes between the top 39.[3] What makes the difference between a winner and all of the others on a particular weekend?"

"This is a great example of alignment. For instance, when Tiger Woods wins, he is aligned with his goal both physically and mentally. When golfers hit difficult shots and high-pressure putts winning tournaments, they are said to be "on." What does that mean? I would like you to consider that they are in alignment. It is as if all of the pins in a combination lock have fallen into place. I'm sure that each of you have experienced this feeling in your life. When you have felt an incredible surge of your personal power during some activity at some point in your life even if it was just for a moment. At that moment you were passionate and in alignment and your results or your feelings displayed that."

There was confirmation in the eyes of the students.

"I must point out that not having all of the pins in alignment does not preclude you from having success in your life, but it is more of a struggle. It is like trying to swim up river instead of going with the flow. And that is what I have been trying to express to you over the past couple of bonfire lessons."

"Don't be afraid to grow and move yourself into alignment. Do not resist the universe and the longing inside of you waiting to be expressed. Connect with the universe by being quiet and through meditation. Ask for guidance and it will come. Then step into the unknown. Open the door in front of you, which presents many more options. Be passionate about your life and your journey. Open the combination lock and you will free yourself, expressing what the universe has been patiently waiting for you to experience."

The crew just sat there absorbing his message as he scooped a pail full of water and threw it on the remaining coals. Carolyn was the first to get up and give Uncle Jack a great big loving hug. The other two followed suit. Tonight they realized that the responsibility of their success in life rested with them. It was easy to set goals, however, overcoming obstacles and challenges and to ultimately live their life in alignment was the key.

Key messages or ideas to take away from this Chapter:

1. Obstacles and challenges are natural to manifestation. They are guideposts directing your path.

2. Negative self talk, frustration, discouragement and peer criticism all sabotage manifestation.

3. Pain and pleasure keep you in your comfort zone. You seek pleasure, but you fear the pain of the effect of positive change.

4. Taking action, conquering obstacles and challenges improves attitude and behaviors and provides confidence to take on more. It is a self-fulfilling prophecy.

5. You must replace a negative behavior with a positive behavior.

6. Quantum mechanics describes a law where there are no pre-defined outcomes. Step into the unknown, it will work itself out, there are many possible outcomes...take action.

7. Be aware of your own self-talk. Recognize when your self-talk is negative and move forward with positive thought and action. You can be negative about everything, or look upon it as a challenge. Understand that this a choice.

8. Life is about education whether it is formal or through practice. Each challenge is a teacher that provides wisdom. Don't turtle. Take risks. Grow.

9. Seek professional help when needed. Recognize this and be honest with yourself. Be positive about this. Remove obstacles and blocks, just as you would remove a tooth if it had decayed.

10. Meditating, fitness and hobbies free the mind to be able to think creatively.

11. Live in alignment with your goals and dreams and the universe will conspire to deliver your goals and dreams with ease.

8 *The Energy Bank Account*

If one advances confidently in the direction of his dreams, and endeavors to live the life which he has imagined, he will meet with a success unexpected in common hours.

—Henry David Thoreau

The afternoon ended with a spectacular display of lightning and thunder. Nature's drama lasted almost an hour and had more climaxes than any action thriller. Every few minutes the dark sky sizzled as lightening strikes lit up the sky like a national firework celebration. The zigzag bolts were of varying colors and iridescent hues. When the bolts were within the vicinity of Pine Ridge, the noise became deafening. The raw power of nature left the residents in awe. Raindrops as big as acorns pelted the deserted beach as the rafts flailed in the now boiling water. The trees bowed to the greatness of Mother Nature as the wind whipped up the cliffs and concentrated her efforts on them. The wind and rain punished all structures in its way tossing window shutters loose causing them to hammer away at the window frames.

And almost as quickly as it came the storm began to subside and all of the residents breathed a sigh of relief. The wind started to die down and the rain sounded like a serenade on the rooftops of the cottages. The storm that had rolled in from the distant west was finally disappearing in the east. Calm had been restored. Eventually the sun peeked from behind the clouds illuminating the sky, just as it was about to dip away for the evening. The last rays of the sun outlined the clouds with a bright white border, and the beautiful colors of a partial rainbow painted the sky.

Activity on the river began very quickly as fishers ventured out on their boats casting their bait in the calm evening waters. Uncle Jack uncovered the wet tarp protecting the firewood and began to arrange nature's fuel for the evening's bonfire. As the sun disappeared and dusk had set in, the bonfire's intense heat began to dry the beach around it. As the smell of the burning wood wafted in the air, it was a signal to return to the beach. As they peered over the cliff, Brian and Tom could see the light of the bonfire reflecting on Carolyn and Uncle Jack and they hurried to join them, knowing that tonight was their final bonfire lesson.

"What an end to the day," exclaimed Tom.

"I thought a few trees around the cottage were coming down for sure," echoed Brian.

Carolyn entered the fray, "some of those lightening strikes were very close to us. I tried to count seconds between the lightening strikes and the thunder claps to see if I could tell how far the strikes were from us. If I remember correctly, each second between the lightening strike and the thunder means that the strike is about a mile away, as sound travels at a much slower speed than light. In a few instances I never even had a chance to count to one. That proves they were too close for my comfort."

"Yes, the energy display we saw today was awesome and quite fortuitously it provided with me with the perfect introduction for our chat this evening, I trust you'll find it interesting," said Uncle Jack.

And with that announcement Uncle Jack started the evening's lesson by posing a question.

"What is energy?"

There was a pause and then Tom said, "Okay, you'd think we would know that by now...but I can't really tell you a precise answer."

"The Webster Encyclopedic dictionary states that it is internal or inherent power; the power of operating; whether exerted or not...and it goes on with other similar words. The reality is that everything is energy. While Webster's focuses on defining energy as a verb as in the ability to do work, *energy is everything*. Einstein taught us that energy transitions into mass at the speed of light squared ($E=mc^2$). However, the real key is that energy and mass are equivalent, as we have discussed over the past few evenings. So energy can be in a form such as light or a rock or the logs we have in the bonfire. Everything in our universe is energy. As we have studied in school, water turns into steam as the water is boiled, as it cools and condenses it turns back into water."

"If everything is energy, then all objects or energy vibrate at different rates. Warm water vibrates at higher rates than cold water. Hot water vibrates at higher rates than warm water. Steam or gas vibrate at higher rates than liquids or solids. The invisible spectrum of light such as x-rays or gamma rays have higher frequencies than the visible spectrum of light. This explains simple things like why warm water cleans better than cold water, because the water molecules in hot water vibrate at a higher rate and thus agitate more effectively."

"That is a great explanation. So what you are saying is that the higher the frequency of the vibration, the less likely that the energy is visible. And in contrast if the frequency of vibration is less we enter into the visible spectrum of light right down to what we call solid objects such as rock. But the key is that they are all energy", said Carolyn.

"You have got it in spades. Great job." said Uncle Jack.

"Furthermore, energy never disappears it just changes form. This is the conservation of energy law. There is a finite amount of energy in the universe which was originally stored within the singularity, the origin of the universe. The initial Big Bang transformed that energy into our universe and everything within it, including us. The energy within the universe is constantly being exchanged. We draw upon that energy in our daily lives. We share in that exchange with the entire universe. We personally consume energy, we create energy, we receive energy and we give energy."

"There was a lot of energy released in the storm today," said Carolyn.

They all nodded in agreement and then Uncle Jack continued. "We receive energy from everything we ingest and through the air that we breathe. We also receive energy from other daily experiences, such as what we see, hear, smell and feel. Think of your own experiences. If they are pleasant and uplifting they energize you, if they are frustrating and exhausting they drain you. People affect you and you have that same effect on them. You may never have thought about this before tonight, but human beings are really conductors of energy."

"I guess you could say that energy flows in and it flows out of you just as electricity flows into any appliance such as a radio, hair dryer or a light bulb. Is that what you are trying to convey?" asked Brian.

"I like your example Brian, picture electricity flowing into a light bulb. If there is plenty of electricity, the bulb burns brightly. If there isn't enough energy the bulb burns dimly. If you managed a hydroponics farm, you would want the bulb to burn brightly. Why? In this case, the light bulb is providing energy for the plant to grow through the process of photosynthesis and as you know, photosynthesis is the key to all plant life."

"The point of all this is to demonstrate that we are conductors of energy. We radiate this energy to everyone and everything around us. This energy can be contagious. A smile to another person is often returned with a smile. A frown may create a different effect. On the receiving side we may not be aware that we pick up energy from everything that is around us. We are affected by what we hear, see, smell and feel. Studies have shown that even the color of a room or the odors that we smell have an effect on our mood. Perhaps you can relate to how you feel when you see pleasant images versus disturbing images. Similarly, listening to pleasant music versus awful music either moves you or grates on you. And of course, love is a powerful energy that moves us in many ways despite the fact that it is invisible other than the visual and auditory clues."

"I'd like to discuss how we exchange energy on a very mechanical level and then look at how this extends to a more energetic level through an example with art. On a mechanical level we are constantly exchanging molecules when we are in a group of people whether in a room, train, car, plane, bus etc. The molecules we breathe are being exchanged with people in our surrounding area. We are constantly exchanging molecules with others. An easy example to draw upon is if we are in a room with a smoker, as they draw on their tobacco and eventually exhale we will eventually draw in their breath, which wafts in the air clinging to the fine particles of smoke they have exhaled. Many of us have experienced this but perhaps you didn't think that you were exchanging molecules. Now consider a room full of smokers."

"This is making me feel a bit queasy," said Carolyn.

"I'm sorry to say that we are in this constant exchange with everything in our environment. We don't need smoke to experience this phenomenon, it happens whether there is smoke or not. Smoke is a great visual tool. And of course in medicine we have proven that many diseases or viruses are spread in tiny airborne droplets that are expelled through coughing, sneezing and breathing."

"I never really thought about how connected we are with other people," said Brian.

"Furthermore we are also in this same exchange with our environment. From the coffee you smell in the morning, the exhaust from the cars ahead of you on the road, to the chlorine at your local swimming pool. The molecules from these substances are entering our bodies and the proof comes from our sense of smell. We live in a large ecosystem we call earth, yet earth is not the container for our ecosystem. Our ecosystem is the universe."

"What do you mean by that? How does our universe become our ecosystem? I realize we are affected by the sun through its energy, this is as clear as can be, but the universe?" asked Tom with a puzzled look on his face.

"Astrology establishes a belief that the stars and planets in our universe affect us. Do you find that hard to believe? If the moon and the sun have such a great effect on our ocean tides, ask yourself, what effect do they have upon us directly? What effect does the moon and sun have on the water within the cells of our body and how does this affect us in a general way? At this point there isn't any type of instrument that can tell us conclusively that there is an effect, but it doesn't take much of an imagination to consider that the moon, sun, and other stars and planets have some effect on us even in a minute way. We do know that large masses create a space-time curvature and produce gravity. What we don't know is how these celestial objects that are thousands and millions of light years away effect us on earth."

"There are other energetic forces that we are aware of. We can all remember spreading iron filings on a piece of paper in our science classes and holding a magnet nearby. The invisible magnetic energy begins to align the iron filings in the direction of the magnetic field. This invisible magnetic field is also present around our planet and it will deflect a compass needle anywhere on earth which indicates the direction to the closest magnetic pole. We call this force or energy the electromagnetic force."

"Also consider this. Every now and then there are solar flares that burst from the sun. These balls of energy hurl through space with great destructive force. The energy is primarily in the electromagnetic spectrum, from gamma rays, x-rays, visible light and radio waves. Fortunately we have some protection on earth due to the atmospheric layer surrounding the planet. In fact it is a protective magnetic field. However, every once in a while these energy bursts knock out communication satellites and power grids such as in Quebec in March, 1989. You may have experienced these events occasionally with some of your communication devices such as radios or cell phones when occasionally solar radio waves overwhelm cell phone towers which cause dropped calls."[1]

"A solar flare spews forth coronal mass ejection. A coronal mass ejection may contain a billion tons of matter that can be accelerated to several million miles per hour and this affects anything in its path. As you know light reaches the earth in about 8 minutes, but the protons from these eruptions can reach earth in as much as 30 minutes with the bulk of the mass taking up to 3-4 days."[2]

"What effect do these bursts have on humans? On earth, we are protected by the atmosphere and magnetosphere, however, at higher altitudes there is less protection. In the month of November 2003, the news media reported that there was such a major solar flare that it created a concern for the astronauts in the International Space Station and for people flying in airplanes due to the extremely high levels of radiation. Some airlines directed their planes to fly at lower altitudes to protect their passengers."

"On some summer nights we see incredible displays of the Aurora Borealis or Northern Lights. This display is created when charged solar particles hit the earth's magnetic field and excite oxygen and nitrogen atoms in the atmosphere. When major solar flares erupt and travel to earth the earth's magnetic field is overwhelmed and the Northern Lights are seen well into the southern regions of the northern hemisphere."[3]

"So there is a lot of energy coming at us from the universe," said Carolyn.

"There is no question about that. Let me give you an example of weak energy that will help cement this point. Voyageur

I has been on an unmanned space mission for 26 years after having been launched in 1977. It is now twice as far as Pluto, traveling at 38,000 miles per hour and is the most distant human made object from earth. It takes approximately 10 hours to communicate with the earth tracking stations as its radio signals travel to earth at the speed of light. The radio signal is produced with power that is 20 billion times smaller than the average digital watch battery.[4] The point is that we can receive and decode Voyageur's weak communication signal on earth, which is millions of miles away. Think about the powerful x-rays, gamma rays and gravitational forces that are produced by the sun and stars in the visible night sky, and what effect they have on us. This is something that astrologers, scientists and philosophers have been exploring for centuries. There is no doubt that there is some effect."

"Now let's consider an artist who paints and how energy is transformed within this realm. In this case energy transforms itself from inspiration in the artists mind, to canvas and then into the mind of the observer," said Uncle Jack.

"Think of the famous painting of the Mona Lisa. Picture her face and think of the color in the picture. Can you see her slight smile on the muted canvas? How many people on this earth can picture the Mona Lisa in their minds? Thousands, millions or perhaps billions of people. Isn't that amazing? Leonardo DaVinci created this masterpiece hundreds of years ago and the energy he used to create this masterpiece goes hundreds of years beyond his lifetime and stretches out to millions upon millions of people."

"That's amazing, I have never seen the real painting, which is on display in the Louvre in Paris and you are right. I can picture it yet I have no idea where I have seen it," commented Tom.

"During the creation of the Mona Lisa, energy was transformed into mass. An inspiration or thought was converted into a painting. The Mona Lisa could be a real person whom he painted or could be a composite of many people that he knew. In the creation of the Mona Lisa, mass which is Mona was converted to energy within the observer's mind. Then it is re-converted back into mass in the form of the Mona Lisa painting. Mona Lisa the real person (M-Mass) came into Leonardo's mind (E-Energy) and then transformed back to art on a canvas (M-Mass). Then it is converted back into energy in the mind of the observer of the painting, or anyone who looks at copies of it in books, views a copy on the internet, posters, movies etc. Even right now, you probably cannot express where you have seen this image of the Mona Lisa that is locked in your mind, it is just there in the form of energy. If you were really artistically inclined you could draw the image from your mind and re-establish it in a material object. Of course it would be your translation of the Mona Lisa, because the likelihood is that it would differ somewhat from DaVinci's version."

"You can bet on that one. I can picture the Mona Lisa too. If I had to paint it I think it would be considered a Picasso," stated Brian, and knowing Brian's painting skills the group had a good giggle.

"If the Mona Lisa is in your mind, what else is? Arguably everything you have ever experienced, all of the visions, thoughts, sounds and smells are locked in there. If you couldn't picture the Mona Lisa as suggested in my earlier example, perhaps the smell of a certain perfume or food reminds you of a certain person or an experience from your past. Perhaps the smell of smoke from burning wood reminds you of hot summer nights and our bonfire chats beside the river."

"That is the truth," said Carolyn.

"Some of these memories, which are simply stored energy, are very deep and may not be on a conscious level. In fact, in Montreal during the 1940's one of the early pioneers of brain surgery, Dr. Wilder Penfield, experienced very interesting results while operating on a patient. During surgery he was able to get his conscious patients to recall memories from the distant past by stimulating brain activity with the use of electrodes. These were memories that they had not consciously remembered."[5]

"That is amazing. I have read about brain surgery and it is true that the patients are typically conscious during their surgery," said Tom.

"That would be a truly weird experience," said Brian.

Uncle Jack continued, "Everything you do on a daily basis transforms energy and you are firmly connected to the universe in this exchange. Da Vinci's Mona Lisa is one very small example of energy transformation and its effects. It is important to understand that you can transform energy in both positive and negative ways. Dwelling on a any particular negative situation can cause you to spiral down a negative energetic path, however, by focusing on the positive energy of any situation can bring dramatically different and more satisfying results."

"What we think about on a day-to-day basis is extremely important. The expression 'what you think is what you are' is very powerful. You have to be consciously aware of the decisions you make daily. Most people are sabotaging themselves without being consciously aware of it. The easiest example of sabotage is through negative self-talk whereby they never even start to manifest their great thoughts or ideas," said Uncle Jack.

"In life, the most important thing you can do for yourself is to surround yourself with positive energy. You are a conductor of energy with all the inputs and outputs that you experience. Not only are we a conductor of energy, we also store energy. Perhaps the easiest metaphor to use to understand this is a bank account. The most obvious measure of energy capacity is related to the need to eat or sleep. If you are low on energy, you need to replenish the energy bank account."

"Perhaps less obviously, everything you experience daily also affects your energy. Positive experiences build up your energy bank account and negative experiences drain your energy bank account. Each small transaction whether a debit or a credit adds up during the day, the weeks, the months and even the years. This is no different from your financial status and financial activities. You either build up capacity energetically or you deplete it. I think you get the picture."

"That is a great example, please go on," prompted Brian.

"The best thing about your own energy is that you are always going to be the manager of it. You get to decide how to build it, store it and use it. However, there is the converse side, if you are not a very good manager, you'll go bankrupt. This is a sad situation. What happens in energy bankruptcy? You become lethargic; perhaps you will eventually compromise your health. You may quit your job, give up on school… the list goes on."

Tom said, "Yes there are times when I have completely drained myself and I am of no use to anyone at work or even to my friends. I feel like I'm exhausted all of the time. I guess you could say that in these cases I'm definitely bankrupt from an energy perspective. Come to think of it, the times when I'm most drained, I seem to come down with a cold."

Uncle Jack continued his talk. "We all manage our energies differently. We are all at different stages of our life's experiences. Some people use their wisdom and strive to keep their lives at a higher level of energy. Others may not have the experience to see the bigger picture and they need to become energy bankrupt in order to recognize that they must make different decisions in life. This is no different from a financial bankruptcy."

"Energy bankruptcy can be viewed in two different ways. It can spell further disaster or it is the calling card for change. Any obstacle you encounter provides an opportunity to learn, it is part of the school of life. By becoming bankrupt energetically or financially, you learn the hard lessons of life. The next time you start to experience similar results you can begin to change your tactics and strategy, otherwise, you are back to learning the lesson over again. We have all had times where we have been 'slow learners' in the school of life."

"That is a great lesson Uncle Jack. This really is making

sense to me. I can now recognize some of the repeating cycles I'm going through with work. I guess I'm a slow learner but if I have the awareness I can break that cycle," stated Tom.

"Eventually you will figure it out. As we discussed last night, challenges and obstacles should not be viewed as negative things or events. They are guideposts for your future. These experiences allow you to change your tactics and strategies. They are opportunities to learn. It's all about perspective isn't it? Life is all about perceptions – reality is different for everyone."

"Perspective is a very individual concept. Everyone comes from a different worldview. The *right answer* is only the *right answer* depending on your perspective. That's not to rule out that a person's perspective will change, because it will with more knowledge. Our worldview evolves."

"Back to energy and energy bankruptcy. I'll give you two examples to illustrate energy bankruptcy and its effects. We often hear about a cycle of life from which people in the poor areas of North America cannot extricate themselves. On the surface it looks like a financial issue but it's really all about energy. Life is not represented by dollars. Money is an indicator of what you are doing in your life. You may be happy and feel successful in many other facets of your life such as in your spirit, your relationships, or your life's adventures. You could be a multi-billionaire but be completely bankrupt with respect to emotional and relationship energy."

"More to the point, the issue in many cases is that too few successful role models exist to assist the poor in realizing that they can break the cycle and make an exciting life for themselves. They are trapped in a paradigm that does not promote an easy exit. There is a tremendous amount of energy in these communities and this capacity is never acted upon. Unfortunately this capacity is not nurtured. Many do escape the poverty and we have heard of tremendous leaders who have come from very impoverished backgrounds."

"Oprah Winfrey is someone who is incredibly successful.

She fits into this mold doesn't she? She was raised by her grandmother on a farm without plumbing and later lived separately with her divorced parents at different times before she went to college," stated Carolyn.

"Yes, it is these types of people who inspire others to break the cycle. They may be teachers, political leaders, religious leaders or private citizens. History presents us with Gandhi and Mother Theresa as two exemplary leaders of modern time," added Uncle Jack.

"The other good example of energy bankruptcy is demonstrated when a person's health has been compromised. As we discussed last night, perhaps they have had a big scare such as a heart attack. Some people wither in this fear while others are inspired to change their lives. They take on a new perspective of their life, perhaps they are inspired to take on a new job or even retire from their job, and begin exercising and radically changing their diets. It becomes an exciting opportunity to regenerate their energy capacities. Therein exists the perfection of life. There is a balance required in life and when the scales tip drastically in one direction or another, an event may come about to bring them back into balance. So, a heart attack or an illness can be a blessing in disguise as long as it isn't fatal."

"You are so right. I know many older people who have started new careers or finally started healthy diets and exercise programs when they were told they had a serious illness to deal with," said Brian.

"Everything happens when there is a deadline. I guess your point is don't get enamored with dead...it's not good," exclaimed Tom.

"Your experiences are extremely important to manage. It starts from birth so don't wait until you get older. The people you surround yourself with, what you read, watch and listen to, all affect you. Don't let other people drain you. If they don't fit into your world, let them go. Fire them. This is a conscious choice you must make. You must make good decisions to be an effective self-manager. Seek out inspiring experiences. Move away from people or situations that want to suck you into their misery and drain your energy bank account."

"I really think I need to fire a couple people that I call friends. They always seem to be trying to better me in everything I do and they play with my emotions. It is like mental abuse that I don't need to endure. I keep fooling myself to think I can change them. They are great when they are positive, but when they are negative it is ugly. I understand what you are saying, I need to better determine with whom I spend my time. There are other real friends that are a pure joy to be with," explained Carolyn.

"That is a very good link Carolyn, I am proud of you. It is difficult to remove negative energy at times, but you must recognize it and do it. To break off the relationship can be very difficult, but once it is done it is over and you go forward feeling lighter and happier," said Uncle Jack.

"Does that mean you avoid controversy completely or avoid assisting people in need?" asked Brian.

"Absolutely not, the satisfaction of managing difficult issues can bring a great deal of energy back into your energy bank account. Be careful with your energy bank account and don't become overdrawn. Just like your day-to-day finances, being in debt consistently is ineffective. You have to manage your energy effectively. Therefore, you must have a strategy to build up your energy."

"How do you do that?" asked Carolyn.

"Fortunately managing your energetic bank account is simple. Intuitively we all *know it*, however, it takes discipline to *achieve it*. Discipline is the difficult component of this simple strategy. However, if you are motivated to change then it's easy. Here is the simple plan."

1. Eat the right fuel to manifest efficient energy. Eat plenty of natural protein, vegetables and fruit. Don't

forget lots of pure water.

2. Exercise. Train your body to handle physical work. We evolved from very physical beings that hunted for food and survived on the land, our ancestors didn't spend time on computers or in offices doing paperwork. By exercising our bodies we release natural stress relieving hormones. Exercising also increases the level of oxygen that goes to our cells which helps optimize the functioning of our physical bodies, which improves our energy levels. The focus meditates our mind to be in the present moment...not in the past or in the future.

3. Get plenty of sleep. Without the proper amount of sleep we become lethargic and error prone. Lack of sleep can compromise your health.

4. Inspirational energy. The best sources of this are from relationships with friends that inspire you, inspirational books, spiritual study, meditation and hobbies that rejuvenate you and that you are passionate about.

5. Remove the negative energies within your life. End relationships with people who drain you. Seek strategies to change other negative areas in your life. Reduce or eliminate caffeine intake, smoking and alcohol.

"Was that plan surprising?" asked Uncle Jack.

"No," they chorused.

"I didn't think that there were any surprises. It's simple, effective and it just makes sense. So why is it so difficult for some people to follow through with this plan? It takes discipline. Unfortunately, until there is a significant reason, most people will not change. All of you are here at this bonfire chat for a reason. You are open to change. The trigger for some people might be a compromising illness or another major event in their lives. My question for you is: *What are you waiting for?*"

"I don't plan on getting sick Uncle Jack. You're right we have to start right now to be aware and have discipline. We must be responsible for ourselves. We control our lives. It is our life to live and our bank account to manage," exclaimed Tom.

"Earlier I said that energy never disappears it just changes form. Let's talk about this subject in more practical terms as it applies to your life. Think about the following expressions:

§ What you put in is what you get out;
§ What you sow is what you reap;
§ What goes around comes around."

"The net of these statements is that whatever amount of energy you put in is what you will get in return. The energy you expend transforms into other energy, which may be a service, or it can be converted into mass, which is a material object. As Einstein postulated, energy and mass are the same thing."

"Fitness can help you channel your energy and can help re-energize you. Sometimes you are not sure if you have the energy to go out for a run after your tiring day in which you perceive you have expended most of your energy. You may think your energy bank account is empty or even worse in overdraft. This is the moment that discipline counts. You may have an internal dialogue wherein you must decide between action or inaction. If your discipline is strong you will get up and put on your running shoes and head out for a run. The motivation needed to get out of the house was tremendous. As you start the run, it seems difficult and you are not sure if you can possibly run for the usual half hour. After laboring for the first 10-15 minutes you still feel like stopping, it's too much work. You question your discipline. As you approach the end, you suddenly start to feel strong, euphoric, like you could run another 20 minutes if you wanted to. When you finish the run and walk for a few blocks to cool down, you definitely feel euphoric. You may decide that life is great and you wonder why you don't do this all the time. You feel like a new person, not at all like the one that had the magnetic attraction to the couch. You may even make a promise to yourself to do it again the next day."

"Isn't it endorphins that are released during exercise that give us the euphoric feeling?" asked Carolyn.

"Yes, you are right. There is a chemical reaction happening in your body. Scientists know that hormones called endorphins are being released within you as you exercise and it gives you that euphoric feeling. It is an addictive feeling. You want to feel it again, but you need to get past the negative self-talk and the couch. Once you equate exercise with positive feelings it becomes easier to sneak past the couch."

"Sometimes in the evening it's tough to get past that couch," said Tom.

"Well it is easier for some people Tom," laughed Uncle Jack, "but isn't it interesting that you expend energy during your day and even more during your run, and yet after your run it is as if you gained energy? You find you are calmer and more able to take on mental or physical tasks. Your body is under the effects of the endorphins, however, your body is also benefiting from an increase in oxygen. The energy stored in the air is transferred into your lungs and blood vessels to the cells in your body. You are optimizing your energy capacity and increasing your energy bank account. Furthermore your focus on running silences your mind from the typical chatter, which results in a meditative calmness, not thinking of the past or the future, just being in your present mind."

"There is nothing like the feeling of expending energy after a hard day's work and then relaxing afterwards," said Tom.

"The quality of the effort expended directly relates to the quality of the result. Practice, rehearsal and studying are one kind of effort. Patience and the skill of a craftsman are yet another. It took Michelangelo many years to paint the Sistine Chapel. His statue of David also took a significant amount of time and energy. He put energy into his works and created masterpieces that have given pleasure to millions of people over time. Not knowing Michelangelo, I can only assume that he enjoyed the pleasure that people received by observing his art. It is what drives most performers, artists, athletes etc. They receive energy back from what they put out. In today's business world, we receive payment for our effort in the workplace in the form of salaries. In our jobs or careers, we trade our energy for money, which we exchange for the food, housing and for the other activities we wish to experience like entertainment, education etc."

Tom popped into the conversation. "This conversation about creating energy and using energy reminds me of the expression, *which came first the chicken or the egg?* This is the age-old question. Did the egg appear first which hatched the first chicken, or did the first chicken appear and lay an egg? The problem is that an egg doesn't appear without a chicken and chicken can't be produced without an egg. Maybe it's easier to ask, *why did the chicken cross the road?*"

With that statement Carolyn, Brian and Uncle Jack burst into laughter.

Once composed, Uncle Jack continued, "The origin of our energy goes back to the singularity. What created it we don't know, however, the singularity provided the burst of energy to create the universe and life as we know it. Like the example of the chicken and the egg, it is a mystery. The most important message to remember from tonight's lesson is that the energy is there for us to manage wisely."

There was a long pause.

Uncle Jack continued, "And here we are at the same point we were seven days ago talking about the origin of our universe, the singularity. We have come full circle. In our nightly discussions we have covered many topics, ideas, theorems and mysteries of life. Each of these topics that took up a small part of our discussions is a lifetime of work for many scientists, researchers, theologians, astronomers, physicists and those that are driven by human dynamics. And, in many cases we are at the tip of iceberg in our collective knowledge. The mysteries that remain propel us as a society to further our research."

"What I hope you have drawn from our discussions is that conclusively we know:

1. Our universe is filled with magic;

2. We all have the ability to use this magic to manifest our dreams."

"Tomorrow night we will have our last bonfire and there I will turn the tables, it will be my turn to listen. Tomorrow I'd like you to share your dreams and goals and some concrete steps that you are going to take to actualize them. I want you to dream and dream big. I want you to define what kind of life you wish to live. Don't hold anything back. Express yourselves; no ideas are too big, too small or silly. I want you to take some steps to start manifesting these goals. A lifetime can go by very quickly, so don't sit on the sidelines watching it go by."

"Each of you will also do an exercise that examines your greatest successes and your life's greatest lessons. We will uncover what you have learned through these experiences and how they have shaped your lives. We will examine what limiting beliefs are holding you back and giving shape to the size and growth of your comfort zone. This energy that you hold from the past does not reflect your future, it has only been a stepping-stone to get you to where you are today. With awareness our paths can be changed."

"Lastly, we will project ourselves into the future and look back to where we are today and give ourselves advice on how to overcome the challenges and obstacles that currently lie before us."

Uncle Jack's eyes were radiant. He was full of life, love and confidence in his bonfire students. This energy was reflected in his students because they also experienced the same. Not a word was spoken as each of the students pondered their future. Uncle Jack extinguished the fire and they headed off to bed with a feeling of excitement and empowerment.

Key messages or ideas to take away from this Chapter:

1. Everything is energy.

2. Energy vibrates at different rates or frequencies i.e. visible light or invisible light like gamma rays or x-rays.

3. The law of conservation of energy – energy never disappears it just changes form.

4. Humans are energy and transformers of energy.

5. We have an energy bank account that we manage personally; similar to a bank account that has a current balance with debit and credit transactions.

6. Awareness on a daily basis is the key to building your energy bank account. Be careful of huge debits without crediting back your account.

7. Don't wait until your back is up against the wall and you have to build your energy bank account. Compromised health is a last resort.

8. Life is in perfection. If the scales tip drastically to one side, then it will rebound back. Too much imbalanced stress leads to compromised health.

9. Remove energy draining relationships from your life.

10. Managing your energy bank account is simple. See the five step plan:
    a. Diet
    b. Exercise
    c. Sleep
    d. Inspiration
    e. Remove negative energies

11. You must be disciplined with respect to your life. Expect the best for yourself, you deserve it.

9 *The Letter*

The intuitive mind is a sacred gift and the rational
mind is a faithful servant. We have created a society
that honors the servant and has forgotten the gift.
—Albert Einstein

It was a perfect September day and Tom was staring vacantly
through the windshield as his minded drifted. He thought about
how he had arrived at this point in his life and realized the debt
of gratitude he owed to his family and friends. With this in mind
his thoughts returned to his destination, Pine Ridge, and to the
purpose of his trip. Tom was attending Uncle Jack's funeral.

The ceremony was to be held on the cliff high above the
river beneath the shade of the towering pines. At a special
bonfire on the evening of the funeral, Tom was to read the
letter. Earlier in the week he had received a telephone call
from Robert Bailey. Robert was a lawyer and an old friend
of Uncle Jack's and had been asked to wind down his af-
fairs. Robert did not know the contents of the letter, he could
only provide the instructions that Uncle Jack had left in his
will. Robert explained to Tom that the group that had come
together for bonfire lessons ten years earlier was invited to
attend a special bonfire on the night of Uncle Jack's funeral.
It was at this gathering that Tom was to read the letter. For
Tom it was a tremendous honor to accept this duty, as he felt
truly blessed to have known Uncle Jack.

Tom was stunned when he heard the news. Upon putting

down the receiver, he felt the hot sting of tears roll down his face as he was overtaken by his emotions. In the telephone call he had learned that Uncle Jack had slipped away silently in his sleep at the age of 89. It was a great loss to all that had known Uncle Jack. Tom was now 41 years old and it occurred to him that he could not remember life without Uncle Jack as he was always at Pine Ridge when Tom was there. Whenever Tom thought about Pine Ridge, gazed at the brilliant stars in the night sky or wondered how he had achieved his goals, he thought about Uncle Jack. He had been his friend, his teacher and his mentor.

Tom began to think about his life and his own mortality and how quickly the years go by. Time seemed to be accelerating. He realized that he was just half of Uncle Jack's age. Tom was overcome with the love he felt for this kind, engaging soul who had inhabited this humble earth in the vast universe for 89 years. The week that Tom had spent in Pine Ridge ten years earlier had changed his life forever and the last ten years of Tom's life had been incredibly rewarding, it was almost as if he had been asleep before, living life unconsciously. Prior to the bonfire lessons, Tom had been living a preprogrammed life. He had been living the way society demanded him to live, ignoring the subtleties of life: such as the magical universe he lived in and how he had the capacity to direct his life as he wished. Fortunately for Tom, Uncle Jack's bonfire lessons had helped him put his life on track.

"How much further?" came the cry from the back seat. The spell was broken and Tom was of present mind. As he looked over his shoulder he saw the adorable round face of Rachel, his 5-year-old daughter and the source of the question who peered back at him from her booster seat.

Without missing a beat, Jamie answered, "We are almost there."

Tom looked over at his wife with loving eyes as he reconnected to the present moment, then he casually wiped away the tears that had streamed down his face as he turned his attention to the road ahead.

Brian had flown to Pine Ridge the day before. He had just come from a meeting with his business partners when he received the message to call Robert. They had just completed the final survey for the new mountain resort in British Columbia. In a few weeks the heavy equipment would be arriving to start clearing the land. As he gazed up at the snow covered mountain tops his eyes filled with tears, thinking back to the days on the beach in Pine Ridge when he shared his dreams with Uncle Jack. How ironic it was that his biggest dream was coming to fruition and Uncle Jack would not be there to see it.

The engines of the plane roared as it sped down the runway at La Guardia. Carolyn squeezed her husband's hand at she thought about their journey to Pine Ridge and the funeral. She had just spoken to Uncle Jack a few weeks before and he had seemed as healthy and happy as always. She recalled how she'd bubbled with excitement as she described the feeling of seeing her first choreographed show debut on Broadway. This was a huge step in her career and she was thrilled to share it with Uncle Jack. Uncle Jack loved to stay in touch and she adored him for it. He had always loved to hear any exciting news she had and he was there to discuss the challenges she faced along the way. Over the years he had come to see her perform in various shows in Toronto, London and in New York. She was going to miss him dearly.

The funeral was a beautiful affair. The smell of the pines wafted in the air as the hot sun radiated in the cloudless sky. The occasional hawk soared in the air circling high above the treetops. There were many familiar faces, some of whom Tom had not seen for many years. He guessed that there were more than 500 people in attendance.

Tom took comfort in the minister's words: "This event is for the living, the friends and family. It is a chance to rejoice in Uncle Jack's life and the energy he imparted to all of us."

The minister continued, "We are all better people because of Uncle Jack. He has left a great impression on our lives. He was a very special person to us and we must remember that every soul is special and they alone make up the fabric of our lives."

A number of Uncle Jack's family and friends spoke at the ceremony and shared their thoughts about him. There was a moment of silence, whereupon the whole group was asked to say a private prayer to this wonderful generous spirit who had departed. At the end of the ceremony Uncle Jack's eldest child, now a grown man, sprinkled his ashes on the cliffs of Pine Ridge underneath the canopy of pines.

As Tom observed the ceremony and heard the words spoken about Uncle Jack he realized that despite the fact that Uncle Jack was not physically present, his energy radiated everywhere. In fact, Uncle Jack's energy was magnified by everyone in attendance, for he had touched them all in some way. As Tom came to realize this thought he smiled to himself knowing that Uncle Jack's energy would never disappear as it had only changed form. A warm sensation spread throughout his body embracing him from the tips of his fingers to the tips of his toes. This was something that Tom had learned at the bonfire lessons years ago. Tom felt this energy everywhere. Uncle Jack's energy remained in the physical surroundings he had created at Pine Ridge; such as his home, the badminton court and in the rafts floating in the river. But more importantly Uncle Jack's energy was in Tom and in the others that were touched by him. They were spreading his message and his teachings about life and the possibilities. They were living *their* lives. This ceremony was a chance for many to publicly express how important Uncle Jack was in their lives and to express their gratitude. Although there were many tears shed that day, there were many more smiles as those in attendance reconnected to the special energy that was Uncle Jack. Tom was not alone in his realization.

The group eventually broke up and either returned to their cottages or back to their homes away from Pine Ridge. Tom, Brian, Carolyn and their family and friends returned to their cottages.

Later in the day there was an unusually large group that collected down at the beach. Screams of laughter came from the river as the children played water games and parents watched with a careful eye while discussions about their lives were shared on the beach. It was a Pine Ridge summer afternoon not unlike thousands that had passed before.

Then, as if on cue, the universe provided a perfect tribute to Uncle Jack's life as a gorgeous Pine Ridge sunset began to take shape. The sun danced behind some clouds near the horizon outlining the dark purple fluffy clouds with brilliant orange before it broke out into a clearing just above the tree line. It cast its warm light one more time before dipping behind the trees. As the sun went down, most of the crowd dispersed to get ready for dinner at Uncle Jack's home. The only exception was for the original bonfire group and their families, for tonight was going to be a special 'by invitation only' bonfire evening.

Tom, Brian and Carolyn had collected wood earlier in the day and piled it beside the bonfire pit. As dusk was upon them they scrounged for more sticks and other dried-out pieces of wood to act as kindling. Tom had the pleasure of starting the fire, taking on Uncle Jack's role.

Sweatshirts, jackets and long pants became necessary and the bonfire was welcome warmth as the cooler September evening was quickly upon them. The children had seats closest to the fire and were toasting their marshmallows in brilliant orange flames and soon the group was singing the old familiar songs, songs that had been sung at many bonfires.

As they finished singing, Robert Bailey handed Tom a manila envelope. As Tom opened it, he pulled out a couple of pages that were hand written and had been addressed to the 'The Bonfire Group' and listed the names: Tom, Brian and Carolyn. There was also a short note of instruction addressed to him. Tom quickly read it to himself. It said, "to be read at Pine Ridge at a beach bonfire on the first clear summer's evening after my ashes have been spread on the cliff. To this bonfire I invite the aforementioned individuals and their immediate families i.e. wives, husbands and children. Tom Armstrong, Carolyn Jones, Brian Smith."

Tom looked up slowly and a hush came over the group, all that could be heard was the crackle of the fire, waves splashing on the beach and frogs croaking in the distance.

"I am honored to be given the opportunity to stand up and speak the words of our friend Uncle Jack, having come together with this group that has gathered today in this uniquely spiritual place. It is with great love and respect that I take on this responsibility and with this understanding I shall focus on the job at hand as we hear his message," said Tom.

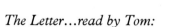

The Letter...read by Tom:

My friends, isn't it funny that you are here at Pine Ridge on a glorious evening sitting around a bonfire listening to me again. (They all laugh and some laugh and cry as the emotion strikes them). I can see you now as I write this. I see your faces and your expressions and I hear your voices and laughter. You are always with me, like DaVinci's Mona Lisa painting, as we discussed years ago. Your energy is always with me and I have always drawn upon it as I wish, like I do at this present moment as I write this letter to you.

I must start by saying thank you. I sincerely appreciate the color you have provided me in my life, for it is different than you think. While you may have thought you were the students, in actuality we were all students learning about the possibilities and capabilities we have as human beings. Your willingness to learn pushed me to learn and to think about the best ways to describe what was complex in a simplistic way. I thank you for the short week we spent together ten years ago and for the relationships we have developed over our lifetimes. I have seen incredible growth in all of you since that week and this growth has been expressed in many ways. I see this in your love of life, your inquisitiveness, the love you share with your new families and special friends, your careers and how you have impacted the people whom you touch in your lives. You have all shared many stories with me over the past ten years and they have all touched my heart. For you see, you have helped manifest my dreams and goals, for I have humbly assisted in providing you with some direction and this is what mattered to me the most.

We have known each other for many years. When we came together to talk about the magic in life in this wonderful place we call Pine Ridge, we talked about the mysteries of life as we sat around a bonfire just as you sit here this evening. You asked questions and I talked and you asked more questions. We took time to go deeper than the trivial aspects of our day. You had the courage to ask questions and discuss subjects that made you think about life and its possibilities.

What did we learn?

We discussed how the world we live in is a magical place, and that the more we learn through science the more spiritual we become in our thinking.

For thousands of years man has connected with the infinite wisdom of the universe through spiritual practices, meditation and prayer. What is prayer? It is nothing more than intention. Today we can see that science and spirituality are one and the same. But were they ever apart or was this separateness just an illusion in the mind of humankind? The ancient philosophers and spiritual leaders taught lessons thousands of years ago that are now being proven through scientific fact. Thoughts, prayers and intention are powerful. In fact, as we know, everything is energy including material objects right through to thought and the concept of soul or spirit.

The phrase *what you see is what you get* doesn't hold true any longer. This is because we don't live in a static universe, it is constantly changing. We can live the life we want to live, we just need to imagine it, design it and live it. Furthermore we all have incredible capabilities to do great things, whether they are large or small. It just takes the courage to take action to start and then to overcome challenges and obstacles on the way.

As our society has evolved many of us have forgotten about the magic that is available to us. Progress has brought us the production line, computer technology and the push to make things faster and for a lesser cost. There are many people who are pursuing their dreams, however, many of us have gotten lost. We have become cogs in the wheel, whether in a factory or an office within corporation. We are doing the things we think we *have* to do to keep up with society.

We are born in a society with many rules. They are unwritten rules, yet they are like a script that the masses follow. At an early age we are locked into a school system that is put together to teach our children, with the best intentions, however, the curriculums are focused on the basics and do not stretch the minds of our youth and give them the options that uniquely inspire them, giving them the capability to reach their full potential. Then after graduation, there are options, either in the work world or in higher education pursuits. Eventually, all of our so called functioning society is off to the world of business that typically runs from 9 a.m. to 5 p.m. five or six days a week, with only one or two days off to pursue their dreams. The job may be the dream initially but eventually it becomes the means to an end. Once this recognition takes place, how are we satisfying our real desires?

In this world we are pre-programmed, domesticated like farm animals from a very early age. How different are we than the early Egyptian slaves who toiled in building the magnificent pyramids in the desert. They were slaves that had to work and in providing their labor they were provided the sustenance to live so that they could work another day. Today much of society works to feed their families and to pay their bills in order to maintain a home to live in. Our work force must work to keep up with society, to maintain their lifestyle and to retain good financial standing. Our leisure time or our time to focus on what we want to do is limited to a few hours a day or on the weekend or during a few weeks of vacation time per year.

It does not conjure up a very pretty picture. Not one that I wish for any of you to ascribe to. We have grown up in a world of perceptions. Reality is what you want, not what you *have* to do. It is as if people have invisible boundaries put around them that keep them locked in a zone of comfort. Imagine an animal being born and raised in a zoo thinking that the cage is the extent of their life, not knowing that their ancestors roamed the Amazon jungle in a territory that in effect is without boundaries when compared to their current home.

This is the point I'm trying to make, your life is without boundaries. You and society have created artificial boundaries that you have adhered to. But the inconceivable is possible and it's waiting for you. You just have to believe in the magic of the universe and manifest your dreams and goals.

I spent much of my time in our concentrated bonfire lessons teaching about basic science. I started with the grand universe we call our home and then delved into the small microscopic world that makes up all material objects. We discussed the theories of relativity and quantum mechanics that govern our lives and how they defy our materialistic perspective of life. When quantum mechanics was first defined it baffled the physicists and had them doubting that our world was as wacky as they were finding out it is. The reactions were much like in Galileo's time when the earth was considered to be the center of the universe. In the past 20 years there are other theories that have been developed such as 'String Theory' that attempt to unify our perspective of nature and attempts to tie together the world that Einstein described in his General Theory of Relativity and that of quantum mechanics.

If I didn't take the time to describe these theories to you and to provide you proof, you may have doubted the magical world that we experience. I'm sure you remember Einstein's thought experiment with the ten year old twins or the double slit experiment that first shed light on the quantum mechanical world. I discussed these theories and concepts that were at times difficult to believe let alone for you to instantly understand their implications in your lives. It was important for me to get our level of understanding of these concepts to a consistent place so that we could evolve to the next step, manifestation.

Through our discussions we learned that we are all energy. Everything we see, touch, do, and experience is energy in one form or another. Every expression you make, every word that you speak or read, imparts energy. I explained to you that Einstein declared that energy is mass, or mass is energy, in his famous equation $E=mc^2$, then I took you to that next step which links your thoughts to energy. That was a simple step as I think we can all conceptualize how thought is energy. However, it was the next link that was the most critical to understand, and that is that thoughts are powerful because they are part of the engine for manifestation to occur.

I talked about how you all are much more powerful than you realize and since that time you have proven it to yourselves. I think today, we can all say that thoughts are very powerful. With each thought, you have the capability to change the world. Just one thought can do that. You can create an object of art that provides joy to you and people who observe or experience it. Just one thought could invent a new material object that could impact a few or millions of people in some beneficial way. Whatever you create whether it is for yourself or for many people produces something new into this world that did not exist before. Isn't that a powerful thought. But you know that already, I'm just reminding you.

As we know, our thoughts are not entirely enough to make powerful changes as you must be courageous and take action for them to appear. You must trust in your instincts and your desire to create whatever you are driven to do. Trust in the magic, for if you apply enough attention to your desires they will appear for you. Why must you be courageous? You must be courageous because challenges and obstacles will appear before you. Some obstacles you will create completely on your own, such as the fear or pessimism that you may not succeed. Expect obstacles and challenges and then figure out a plan to go around them to eliminate them. Never lose sight of your intentions because over time you will realize your goals.

Here are my final thoughts, in summary:

§ We live in a world of magic;
§ Thoughts are extremely powerful;
§ Take action;
§ Expect challenges and obstacles;
§ Re-focus on your intention and take action;
§ The more aligned you are with your goals the more the likelihood of achieving those goals;
§ Never give up learning: read books, take courses, study what interests you and the universe will open up paths that were never there for you before.

As you pursue your goals and dreams, be humble, treat others as you expect them to treat you because as you have heard for years; what goes around comes around. So give gratitude and love to all of the people you come in contact with.

With all my love,
Uncle Jack

With a deep sense of love, they quietly looked at one another acknowledging the special bond they shared that was fostered by Uncle Jack. With that, Tom quietly grabbed an empty pail at the edge of the cliff and filled it with water at the rivers edge. He walked up to the fire and looked in the eyes of those huddled around and with their unspoken approval he extinguished the fire. Hand in hand they gazed at the stars, hugged their loved ones and left the beach together.

Appendix

Goals & Dreams

This exercise has been put together to help you on your manifestation journey. It will allow you dump the swirling thoughts in your mind onto paper in an organized fashion. Please use these exercises as guidelines. You can push further with your own ideas.

The first exercise focuses on how you define yourself and what type of life you wish to live. It challenges you to think of your overall goals and then more specific goals in different areas of your life. Please note that if there are other areas of your life you would like to develop, please add them. Don't limit your thinking by this general format. I have indicated where you can specify other areas. Also, don't adhere to the number of items per section, they are minimums, add as many as you can.

The second exercise focuses on the past. In this space you can articulate what have been your biggest successes and biggest life lessons. For each you will need to explain what you have learned through these experiences and what gifts came as a result. Lastly, list your individual blocks/obstacles/challenges for each goal area and what your actions are to overcome them to reach the goals defined in Exercise #1.

Exercise #1

Who are you? (What would your epitaph say?)

What are your top goals and dreams?

What are your goals in the following life areas (do not limit your goals to 3 in each area...push yourself further):

Relationships/Family

    1.

    2.

    3.

Spirituality

    1.

    2.

    3.

Financial

    1.

    2.

    3.

Health

    1.

    2.

    3.

Diet

    1.

    2.

    3.

Fitness

    1.

    2.

    3.

Career

    1.

    2.

    3.

Leisure/Travel

    1.

    2.

    3.

Creative Self Expression

    1.

    2.

    3.

Personal Growth/Education

    1.

    2.

    3.

Other specific life areas...you define this.

Exercise #2

What are your greatest life accomplishments?

1.

2.

3.

4.

5.

What are your greatest life lessons (where times were difficult and you may have felt you failed)?

1.

2.

3.

4.

5.

What did you learn from each of these successes and difficult life lessons? How have they changed you?

In success area:

1.

2.

3.

4.

5.

In Life Lessons area:

1.

2.

3.

4.

5.

Looking forward, what are the biggest blocks/obstacles/challenges in your life that are preventing you from living the life you wish to live (as you described in exercise #1)?

1.

2.

3.

4.

5.

Before answering the next question please do the following exercise. Take 10-30 minutes and go to a quiet location and quiet your mind. Don't think of the past or the future. Be with yourself, feel the sensations in your body, concentrate on your breathing and free your mind – be present – in the NOW. This step is mandatory.

When you have completed the exercise and feel relaxed... Imagine yourself in the future. Imagine yourself having reached your goals and living the life you wish to live. You are your epitaph and you have experienced your goals. In order to accomplish this, you were able to go past the blocks/obstacles/challenges that you described above. Now answer the next question.

What does the "future you" tell the "present you" regarding these blocks/obstacles/challenges (described above) and how to overcome them?

1.

2.

3.

4.

5.

What gifts have come to you through these blocks/obstacles/challenges?

1.

2.

3.

4.

5.

Now, take what you have worked on to put together actions for each block/obstacles/challenge for each life area goal.

What are your obstacles/challenges in the following areas:	What are your actions to overcome these blocks? When will you act on them?
Relationships	
1.	1.
	2.
	3.
2.	1.
	2.
	3.

3.   1.
                            2.
                            3.

Spirituality
  1.   1.
                            2.
                            3.

  2.   1.
                            2.
                            3.

  3.   1.
                            2.
                            3.

Financial
  1.   1.
                            2.
                            3.

  2.
                            1.
                            2.
                            3.

  3.   1.
                            2.
                            3.

Health
  1.   1.
                            2.
                            3.

  2.   1.
                            2.
                            3.

  3.   1.
                            2.
                            3.

Diet
  1.   1.

2.
3.

2.   1.
2.
3.

3.   1.
2.
3.

Fitness
1.   1.
2.
3.

2.   1.
2.
3.

3.   1.
2.
3.

Career
1.   1.
2.
3.

2.   1.
2.
3.

3.   1.
2.
3.

Creative Self Expression
1.   1.
2.
3.

2.   1.
2.
3.

3.   1.

                                    2.
                                    3.

Personal Growth/Education
    1.   1.
                                    2.
                                    3.
    2.   1.
                                    2.
                                    3.
    3.   1.
                                    2.
                                    3.

Other...you specify

Notes

As indicated below, the following authors inspired many of the facts found in this book. Each author leverages work from others before them and the journey continues. I encourage you to research these materials on your own.

Chapter 2

1. Hawking, S. 1996. A brief history of time. Canada: Bantam Book, page 52

2. Ibid, page 84-85

3. Ibid, page 52

4. Ibid, page 86-87

5. Ibid, page 53

6. Ibid, page 97

7. Ibid, page 98

8. Ibid, page 9

9. Ibid, page 38-41

10. Ibid, page 9

11. NASA Web Site: Cosmology 101: Big Bang Concepts: *map.gsfc.nasa.gov*

12. NASA Web Site: Cosmology 101: Big Bang Concepts: *map.gsfc.nasa.gov*

13. Hawking, S. 1996. A brief history of time. Canada: Bantam Book, pages 123-124

14. Ibid, page 85

15. Ibid, page 124

16. Dickinson, T. 2001. Nightwatch – A practical guide to viewing the universe. Willowdale: Firefly Books Ltd., page 20

17. Hawking, S. 1996. A brief history of time. Canada: Bantam Book, page 19

18. Dickinson, T. 2001. Nightwatch – A practical guide to viewing the universe. Willowdale: Firefly Books Ltd., page 16

19. Ibid, page 24

20. Ibid, page 24

21. Ibid, page 16-20

22. Ibid, page 23

23. Ibid, page 18

24. Ibid, page 23

25. Hawking, S. 1996. A brief history of time. Canada: Bantam Book, page 189

26. Ibid, page 174

Chapter 3

1. Gribbin, J. 1998. Q is for Quantum – Particle physics from a to z. New York: The Free Press, page 52

2. Hawking, S. 1996. A brief history of time. Canada: Bantam Book, page 12

3. Bodanis, D. 2001. E=mc² – A biography of the world's most famous equation. Canada: Anchor Canada, page 97

4. Capra, Fritjof, 1983. The Turning Point: Science, Society, and the Rising Culture. New York: Bantam Books, page 76-77

5. Gribbin, J. 1998. Q is for Quantum – Particle physics from a to z. New York: The Free Press, pages 224-225

6. Ibid, page 173

7. Hawking, S. 1996. A brief history of time. Canada: Bantam Book, pages 55-59

8. Capra, Fritjof, 1983. The Turning Point: Science, Society, and the Rising Culture. New York: Bantam Books, pages 76-77

9. Ibid, pages 76-77

10. Wheatley, M. J. 1992. Leadership and the new sciences. San Francisco: Berrett-Koehler, page 11

11. Chopra, D. 1993. The seven spiritual laws of success. San Rafael: Amber-Allen Publishing and New World Library, page 69

12. Gribbin, J. 1998. Q is for Quantum – Particle physics from a to z. New York: The Free Press, pages 109-113

13. Ibid, pages 109-113

14. Ibid, pages 109-113

15. Ibid, pages 109-113

16. Ibid, pages 109-113

17. Ibid, page 320

18. Ibid, page 258

19. Woolf, Vernon1990. Holodynamics. United States, pages 25-26

20. Woolf, Vernon1990. Holodynamics. United States, pages 25-26

21. Woolf, Vernon1990. Holodynamics. United States, pages 25-26

22. Gribbin, J. 1998. Q is for Quantum – Particle physics from a to z. New York: The Free Press, page 258

23. Capra, Fritjof, 1983. The Turning Point: Science, Society, and the Rising Culture. New York: Bantam Books, pages 81-82

24. Gribbin, J. 1998. Q is for Quantum – Particle physics from a to z. New York: The Free Press, page 322

Chapter 4

1. Bodanis, D. 2001. $E=mc^2$ – A biography of the world's most famous equation. Canada: Anchor Canada, page 69

2. Ibid, pages 80-82

3. Ibid, page 82

4. Ibid, page 52

5. Hawking, S. 1996. A brief History of Time. Canada: Bantam Book, page 30

6. Bodanis, D. 2001. $E=mc^2$ – A biography of the world's most famous equation. Canada: Anchor Canada, page 52

7. www.nasaexplorers.com — The Theory of Relativity and Time Dilation

8. www.nasaexplorers.com — The Theory of Relativity and Time Dilation

9. www.nasaexplorers.com — The Theory of Relativity and Time Dilation

10. Hawking, S. 1996. A brief History of Time. Canada: Bantam Book, pages 30-32.

11. Ibid, pages 30-32

12. Ibid, page 33

13. Ibid, pages 33-44

14. Bodanis, D. 2001. E=mc^2 – A biography of the world's most famous equation. Canada: Anchor Canada, pages 83-84

15. Hawking, S. 1996. A brief History of Time. Canada: Bantam Book, page 34

16. Ibid, page 164

17. Bodanis, D. 2001. E=mc^2 – A biography of the world's most famous equation. Canada: Anchor Canada, page 76-77

18. Ibid, page 112

Chapter 5

1. Web site for the Sudbury Neutrino Observatory: www.sno.phy.queensu.ca

Chapter 6

1. Chopra, D. 1993. The seven spiritual laws of success. San Rafael Amber-Allen Publishing and New World

Library, page 68

2. Ibid, pages 67-69

Chapter 7

1. Corcoran, B. 2003. Use what you've got & other business lessons I learned from my mom. New York: Penquin Group, pages 119-121

2. Ibid, pages 119-121

3. www.pgatour.com — PGA Tour 2003 scoring average

Chapter 8

1. www.space.com — Major Solar Storm Begins Onslaught of Earth October 28th, 2003 Robert Roy Britt

2. edmall.gsfc.nasa.gov/99invest.site/ace/answerkey.html

3. www.space.com — Major Solar Storm Begins Onslaught of Earth October 28th, 2003 Robert Roy Britt

4. voyager.jpl.nasa.gov/news/pressrelease2.html Press Release 2 — February 13th, 1998. Voyager 1 Now Most Distant Human-Made Object in Space (last updated January 14, 2003)

5. www.pbs.org/wgbh/aso/databank/bioindex.html — People and Discoveries — Wilder Penfield

ISBN 1-41204104-X